I0020133

Logics for New-Generation AI
First International Workshop
18-20 June 2021, Hangzhou

Volume 1
Proceedings of the First International Workshop, Hangzhou, 2021
Beishui Liao, Jieting Luo and Leendert van der Torre, eds

Logics for New-Generation AI
First International Workshop
18-20 June 2021, Hangzhou

Edited by

Beishui Liao
Jieting Luo
Leendert van der Torre

© Individual author and College Publications 2021
All rights reserved.

ISBN 978-1-84890-373-9

College Publications, London
Scientific Director: Dov Gabbay
Managing Director: Jane Spurr

http://www.collegepublications.co.uk

Original cover design by Laraine Welch

All rights reserved.   No part of this publication may be reproduced, stored in a retrieval system or transmitted in any form, or by any means, electronic, mechanical, photocopying, recording or otherwise without prior permission, in writing, from the publisher.

Preface

In recent years, a new generation of artificial intelligence is mainly driven by big data and machine learning techniques, while logic has played little role. This trend will be changed in the development of several new directions of AI, including explainable AI, ethical AI and knowledge-based AI, which correspond to three important directions of logical research: causal reasoning, norms and value reasoning, and knowledge graph reasoning. In an open, dynamic and real environment, to support rational decision making and human-friendly communication and explanation, there are two main challenges in modeling such kinds of reasoning. First, how to deal with information that is typically incomplete, uncertain, dynamic and conflicting? Second, how to effectively explain the results and procedures of reasoning to ordinary human beings? Driven by these research questions, a project titled "Research on Logics for New Generation Artificial Intelligence" (2021-2025) was granted by the National Social Science Foundation of China in 2020, as a national major project. To facilitate the efficient communication and collaboration between members of the project as well as other researchers who are interested in the topics of this project, we decided to organize annual international workshops on these topics. As a result, the First International Workshop on Logics for New-Generation Artificial Intelligence (LNGAI 2021) takes place in Hangzhou, China, 18-20 June 2021.

In this workshop, we received 12 submissions. After rigorous peer-review by the international program committee, 8 long papers and 3 extended abstracts are accepted and included in this volume of proceedings. In addition, 1 long paper and 1 extended abstract from invited speakers of the workshop and 1 long paper from subproject leaders are also included. These papers reflect very well the state-of-the-art of the research orientated to the above two questions.

On one hand, concerning logic foundations for reasoning about incomplete, uncertain, dynamic and conflicting information, new progress has been made in the directions of non-monotonic logics and formal argumentation. Huimin Dong and Yi Wang propose a default modal logic for defeasible reasoning by modeling defaults using the notions of consistency and preference. Dov Gabbay and Timotheus Kampik formalise the Shkop approach to conflict resolution in formal argumentation. Gabriella Pigozzi analyzes arguments in public spaces during the Covid-19 pandemic with theories that have been developed in the

i

literature on argumentation. Chonghui Li and Beishui Liao identify intrinsic and extrinsic argument strengths in collective argumentation and propose mechanisms for two argument strengths to interplay with each other. Zhe Yu and Shier Ju introduce a context-based argumentation framework that allows us to obtain consensus in the multi-agent setting. Bin Wei analyzes plausible reasoning in the context of argumentation to characterize the dynamic changes of the plausibility. Lisha Qiao et al formalize different kinds of collaboration in abstract agent argumentation. David Fuenmayor and Alexander Steen propose an approach of analyzing argumentation frameworks and their semantics based on an encoding into extensional type theory (classical higher-order logic). Among these contributions, argumentation-based approaches are not only able to reason about incomplete, uncertain and conflicting information, but also with good potentials to support human-friendly explanation.

On the other hand, about the logical models and algorithms for reasoning in new generation AI, there are three contributions. David Streit presents a framework for logical experimentation in Isabelle/HOL where STIT logic is embedded. He then proposes a way to use computational tools to automatically check how different ways to define notions of causal responsibility behave in various cases. Heng Zheng and Davide Grossi introduce a formal approach about the comparison of cases in case-based reasoning. Luca Pasetto, et al propose a methodology to translation of contract texts into Defeasible Deontic Logic. Ramit Das and R Ramanujam propose a modal logic for reasoning about strategies in social network games, where players are connected by a social network graph. Réka Markovich and Olivier Roy show six possible formalizations of the right to know and study their logical behaviors.

We would like to thank the authors for their contributions to the workshop and the program committee (Michael Anderson, Katie Atkinson, Pietro Baroni, Christoph Benzmüller, Jianhua Dai, Huimin Dong, Xinguo Dun, Mehdi Dastani, Dragan Doder, Réka Markovich, John-Jules Meyer, Henry Prakken, Ram Ramanujam, Tjitze Rienstra, Olivier Roy, Guillermo Simari, Chenwei Shi, Yì N. Wáng, Bin Wei and Zhiyong Feng) for their careful reviews of the submissions. We finally acknowledge the financial support on LNGAI 2021 from the national key project of Research on Logics for New Generation Artificial Intelligence, Zhejiang University.

Beishui Liao, Jieting Luo & Leendert van der Torre
Zhejiang University, Hangzhou, China
University of Luxembourg, Luxembourg
June 3, 2021

Contents

Intrinsic and Extrinsic Argument Strengths in Collective Argumentation

Chonghui Li[a],[1] Beishui Liao[a],

[a]Institute of Logic and Cognition, Zhejiang University, China

Abstract

In the area of collective argumentation, there might be more than one factor affecting argument strength. In the process of framework merging, weights may be attached to attack relations. In the meantime, agents may have preferences over values promoted by arguments. These two conditions provide different sources of argument strengths. In this paper we identify intrinsic and extrinsic argument strengths in collective argumentation and propose mechanisms for two argument strengths to interplay with each other. Furthermore, we find some interesting relations between these mechanisms and propose some properties to evaluate the collective outcome which is jointly influenced by them.

Keywords: Argument strength, Collective argumentation, Mechanism of interplay.

1 Introduction

In the field of multi-agent systems, formal argumentation is usually used to model either dialogue-based interaction aimed at resolving conflicts of opinions [27,3,23] or the process of game-theoretical strategy selection [30,29,28]. The common merit is to form a unified argumentation framework for all agents and obtain the outcome by means of argumentative semantic reasoning. However, there are other possible scenarios in mutli-agent systems, in which agents have respective observed information and reasoning knowledge. Take the scenario of a smart court for an example. Given the information about a case and based on personal background, each member of the jury may have his or her own comprehension and judgement. Thus he or she would represent his or her comprehended knowledge in the form of an individual argumentation framework and reason independently. In Bodanza's survey [8], modelling this type of scenario based on argumentation is regarded as *Collective Argumentation.*

Currently, there are two main research directions in collective argumentation. One is framework merging, which aims at forming representative collective frameworks and afterwards obtain collective reasoning outcomes [14,13,19,16].

---

[1] lisabell@zju.edu.cn

The other is semantic judgment aggregation, which focuses on obtaining collective reasoning outcomes directly [12,9,10]. In the vein of framework merging, distance-based approach [14] and numerical approaches [16,19,13] tackle the problem differently. Based on all individual frameworks, the former qualifies relations between any pair of arguments into three situations, namely attack, non-attack and ignorance, while numerical approach adopts a quantitative way to treat the disagreement on attack relation. In numerical approaches, the votes for the appearance of an attack relation in individual frameworks are represented as *weight*, from which the arguments acquire a kind of strength in the reasoning stage. Considering that this kind of argument strength is inherently obtained from framework merging, in this paper we call it *intrinsic* argument strength.

Preference plays an important role in argumentative reasoning. In preference-based argumentation framework (PAF) [2], the information of preference might have influences on attack relations [2,1] or reasoning outcomes [21] or both[5]. The value-based argumentation framework (VAF) attempts to provide a formal basis for PAF, tracing the preference over arguments to an ordering over values which arguments promote [6]. In the smart court scenario, multiple jurors may have different attitudes towards value distribution and value ordering promoted by the arguments. These are also factors influencing the strength of arguments and the outcome in the collective framework. In this paper we regard them as a source of *extrinsic* argument strength. We consider a simple situation, assuming the group of agents has a preliminary agreement on value distribution over the common set of arguments and has formed a partial strict order over values as collective preference. Without the loss of generality, an argument is allowed to promote multiple values and a value can be shared by arguments.

Based on above discussions, these two strengths have different natures so that they can not be compared or aggregated. If we ignore either of them, it is probable that a bias lies in the collective outcome. Therefore, we need to design interplayed mechanisms for them to balance. Specifically, we are interested in when intrinsic argument strength is naturally yielded in the process of framework merging, how it interplays with extrinsic argument strength which acts as a given information, and furthermore, how they jointly influence the outcomes of collective argumentation. In Delobelle's paper [16], intrinsic argument strength is used to select the best extension(s). However, if extrinsic argument strength is involved, the influence from it should be included in the result too. In the paper we design interplayed mechanisms, called α-*precedence* to dialectically balance the outputs of these two kinds of argument strengths. The α-preceded mechanisms have some interesting relations between each other and are evaluated with some properties related with set cardinality and social rationality, which verify the aspect whether the mechanisms can diminish the choices for the group and whether the collective outcome exists as a non-empty and unique solution.

The paper is organized as follows. The next section recalls the basic notions of abstract argumentation, value-based argumentation framework and one of numerical approaches in framework merging. Next we introduce two kinds of argument strengths. And then we propose the mechanisms of interplay aimed at obtaining a reasonable collective outcome in argumentative reasoning and evaluate them with some properties. Finally, we conclude the paper by illustrating related work and future work.

2 Preliminaries

First, let's recall some key elements of abstract argumentation frameworks as proposed by Dung in [17].

Definition 2.1 An abstract argumentation framework (AF) is a pair $\mathcal{F} = (\mathcal{A}, \mathcal{R})$ where $\mathcal{A}$ is a set of arguments and $\mathcal{R} \subseteq \mathcal{A} \times \mathcal{A}$ an attack relation over $\mathcal{A}$. We denote $Arg(\mathcal{F}) = \mathcal{A}, Att(\mathcal{F}) = \mathcal{R}$.

The key problem is to determine the sets of arguments that can be accepted together. According to some criteria, a set of accepted arguments is called an *extension*. Let us first introduce two basic criteria: conflict-freeness and acceptability.

Definition 2.2 Given an AF $\mathcal{F} = (\mathcal{A}, \mathcal{R})$ and a set of arguments $S \subseteq \mathcal{A}$, we say that S is *conflict-free* iff $\nexists A, B \in S$ such that $(A, B) \in \mathcal{R}$. We say that an argument $A \in \mathcal{A}$ is *acceptable* w.r.t. S iff $\forall B \in \mathcal{A}$, if $(B, A) \in \mathcal{R}$ then $\exists C \in S$ such that $(C, B) \in \mathcal{R}$.

A set of arguments S is *admissible* when it is conflict-free and each argument in $\mathcal{A}$ is acceptable w.r.t. S. Several semantics have been proposed based on admissible sets. In this paper, we only focus on the standard semantics defined in [17].

- S is a *complete* extension of $\mathcal{F}$ iff it is admissible and each argument acceptable w.r.t. S belongs to S,
- S is a *preferred* extension of $\mathcal{F}$ iff it is a maximal(w.r.t. set inclusion) complete extension of $\mathcal{F}$,
- S is a *grounded* extension of $\mathcal{F}$ iff it is the minimal (w.r.t. set inclusion) complete extension of $\mathcal{F}$,
- S is a *stable* extension of $\mathcal{F}$ iff it is conflict-free and it attacks all the arguments that do not belong to S.

We denote $\mathcal{E}_\sigma(\mathcal{F})$ the set of extensions of $\mathcal{F}$ for the semantics $\sigma \in \{\mathbf{co}(mplete), \mathbf{pr}(eferred), \mathbf{gr}(ounded), \mathbf{st}(able)\}$.

If each agent has his own knowledge representation as an individual framework, one of numerical approaches for framework merging is to combine the profile of individual frameworks to a collective weighted argumentation framework (c-WAF), which has been proposed in [16].

Definition 2.3 Let $\hat{\mathcal{F}} = (\mathcal{F}_1, \ldots, \mathcal{F}_n)$ be a profile of individual abstract argumentation frameworks, a c-WAF is a tuple $W = (\mathcal{A}, \mathcal{R}, w)$, where:

- $\mathcal{A} = \bigcup\limits_{i=1}^{n} Arg(\mathcal{F}_i); \ \mathcal{R} = \bigcup\limits_{i=1}^{n} Att(\mathcal{F}_i);$

- $w(A, B) = |\{\mathcal{F}_i \in \hat{\mathcal{F}} : (A, B) \in Att(\mathcal{F}_i)\}|.$

Example 2.4 Given a profile $\hat{\mathcal{F}}_1 = (\mathcal{F}_1, \mathcal{F}_2, \mathcal{F}_3)$ as shown in Figure 1. According to Definition 2.3, we can obtain the collective weighted argumentation framework W_1 for this profile in the right part of Figure 1.

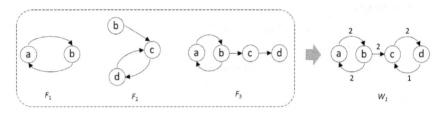

Fig. 1. A profile $\hat{\mathcal{F}}_1$ and the corresponding W_1

When preference is introduced to an abstract argumentation framework, Bench-Capon initially proposed a value-based argumentation framework (VAF) [6], which allows to compare the strength of abstract arguments without referring to their internal structure. Souhila Kaci generalized Bench-Capon's framework to a new value-based argumentation framework [20], in which a value may be promoted by multiple arguments. We use the notation m-VAF to distinguish the class of Bench-Capon's framework.

Definition 2.5 Kaci's value-based argumentation framework (m-VAF) is a 5-tuple $C_m = (\mathcal{A}, \mathcal{R}, V, \mathcal{M}, >_v)$, where $\mathcal{A}$ is a set of arguments, $\mathcal{R} \subseteq \mathcal{A} \times \mathcal{A}$ is an attack relation, V is a set of values, $\mathcal{M}$ is a mapping from V to $2^{\mathcal{A}}$ such that $\mathcal{M}(v)$ is the set of arguments promoting the value v and $>_v$ is a strict total or partial order over V.

Actually, in Bench-Capon's VAF, the value distribution function which maps $\mathcal{A}$ to 2^V is the inverse of $\mathcal{M}$ in Kaci's m-VAF, which can be denoted as $\mathcal{M}^{-1}$.

3 Argument Strengths and Interplay Mechanisms in Collective Argumentation

In our setting of collective argumentation, arguments may acquire different kinds of strengths in at least two stages. In the first stage, through certain kind of deliberation, multiple agents might have a preliminary agreement on the set of values each argument promote and a strict partial order over a subset of values (leave the subset of values which has disputed orderings among agents incompatible[2]), which can be regarded as the source of extrinsic argument strength. In the second stage, the weights associated with attack relations

---

[2] It also can be regarded as a simple form of individual preferences aggregation.

4

during framework merging can be regarded as the source of intrinsic argument strength. Both of them can be regarded as forms of preferences over arguments.

As a qualitative argument strength, preference can influence argumentative reasoning outcomes in two different ways [4]. The first is by modifying the framework. The preference-dependent attack relations would be removed for the reason that the strength of the attacked argument exceeds its attacker's. In Leila Amgoud's paper [5], this kind of attack relation is reversed instead of removing. The other influence is the refinement of extensions. Multiple extensions (if they exist) are compared according to the preference over arguments and the one(s) having the strongest strength are selected. Souhila Kaci'research [21] addresses this.

In our research, on one hand, the collected weights on attack relations can't be applied as the modification of the collective framework, for the collective framework is not a real framework of knowledge representation but somehow a measure of disagreement on attack relations among agents. On the other hand, value information (value distribution and a partial order over values) appears to be common knowledge after the deliberation. We have two options of how to apply it: on individual level or on collective level. If we apply it on individual level, no matter deletion or reversion on attack relations according to value information, some information would be lost or altered from individual level to collective level. Considering that we need to integrate intact information by means of framework merging and study the influences of intrinsic and extrinsic argument strengths on the same level, we choose to apply it on collective level. However, value preference and weights on attack relations have different natures and are incomparable. They can be applied neither together nor respectively as modifications of the collective framework. Therefore, we treat the argument strengths acquired from collected weights on attack relations and deliberated value preference as an interplayed function in the refinement of extensions and design mechanisms to select collective extensions as a reasonable outcome. Here, a *reasonable* outcome for the group is expected to well-balanced between two argument strengths and satisfies some rational properties.

3.1 Intrinsic Argument Strength

First, let's clarify the definitions related to intrinsic argument strength and present with a few examples. We adapt some notions from Delobelle's work[16], such as $\oplus$-*attack* and *best extensions* as *intrinsically stronger* and a *ranking* over extensions.

Definition 3.1 Given a c-WAF $W = (\mathcal{A}, \mathcal{R}, w)$ as Definition 2.3, Let $\mathcal{F}_W = (\mathcal{A}, \mathcal{R})$ be the AF disregarding weights on attack relation. Let E_1 and E_2 be two extensions of $\mathcal{F}_W$ for a $\sigma \in \{co, pr, st\}$. The *intrinsic weight* from E_1 to E_2 is defined as: $\mathcal{E}_{in}(E_1, E_2) = \sum_{\forall A \in E_1, \forall B \in E_2} w(A, B)$.

Definition 3.2 Given a c-WAF $W = (\mathcal{A}, \mathcal{R}, w)$ and its AF $\mathcal{F}_W = (\mathcal{A}, \mathcal{R})$, for any $E_i, E_j \in \mathcal{E}_\sigma(\mathcal{F}_W)$ where $\sigma \in \{co, pr, st\}$, we say E_i is *intrinsically stronger* than E_j iff $\mathcal{E}_{in}(E_i, E_j) > \mathcal{E}_{in}(E_j, E_i)$, denote $E_i > E_j$, and denote the *equality of strength* as $E_i \approx E_j$ when $\mathcal{E}_{in}(E_i, E_j) = \mathcal{E}_{in}(E_j, E_i)$.

Example 3.3 Continue with Example 2.4. Let $\sigma = pr$, the preferred extensions of W_1 are $E_1 = \{a, c\}, E_2 = \{a, d\}, E_3 = \{b, d\}$. Take E_1, E_2 for an illustration. According to Definitions 3.1 and 3.2, $\mathcal{E}_{in}(E_1, E_2) = 2, \mathcal{E}_{in}(E_2, E_1) = 1$, thus $E_1 \triangleright E_2$, E_1 is intrinsically stronger than E_2.

Definition 3.4 Given a c-WAF $W = (\mathcal{A}, \mathcal{R}, w)$, its AF $\mathcal{F}_W = (\mathcal{A}, \mathcal{R})$ and $\mathcal{E}_\sigma(\mathcal{F}_W)$ where $\sigma \in \{co, pr, st\}$, assume the cardinality of $\mathcal{E}_\sigma(\mathcal{F}_W)$ is k, then the intrinsic-weight matrix $\mathbf{A}_{in}$ with square size k can be represented as $a_{ij} = \mathcal{E}_{in}(E_i, E_j)$, where $i \neq j$ and $1 \leq i, j \leq k$.

Example 3.5 Continue with Example 3.3. According to Definition 3.4, we obtain the intrinsic-weight matrix $\mathbf{A}_{in}$ for W_1 as Table 1 shows:

Table 1

$\mathcal{E}_{in}$	E_1	E_2	E_3
E_1	-	2	4
E_2	1	-	2
E_3	5	2	-

Definition 3.6 Given a c-WAF $W = (\mathcal{A}, \mathcal{R}, w)$ and its AF $\mathcal{F}_W = (\mathcal{A}, \mathcal{R})$, the *intrinsic strength score* for an extension $E_i \in \mathcal{E}_\sigma(\mathcal{F}_W)$, where $\sigma \in \{co, pr, st\}$, with Copeland measurement is defined as: $scor_{in}(E_i) = |\{E_j \in \mathcal{E}_\sigma(\mathcal{F}_W) : E_i \triangleright E_j\}| - |\{E_j \in \mathcal{E}_\sigma(\mathcal{F}_W) : E_j \triangleright E_i\}|$.

Definition 3.7 Given a c-WAF $W = (\mathcal{A}, \mathcal{R}, w)$ and its AF $\mathcal{F}_W = (\mathcal{A}, \mathcal{R})$, for $E_i, E_j \in \mathcal{E}_\sigma(\mathcal{F}_W)$, where $\sigma \in \{co, pr, st\}$, the *ranking* over the extensions based on intrinsic argument strength is a total pre-order $\succcurlyeq_{in}$ such that $E_i \succ_{in} E_j$ iff $scor_{in}(E_i) > scor_{in}(E_j)$ and $E_i \sim_{in} E_j$ iff $scor_{in}(E_i) = scor_{in}(E_j)$.

Example 3.8 Continue with Example 3.5. According to Definition 3.6 and 3.7, we have $scor_{in}(E_1) = 0, scor_{in}(E_2) = -1, scor_{in}(E_3) = 1$. Hence the ranking for the extension selection is $E_3 \succ_{in} E_1 \succ_{in} E_2$.

3.2 Extrinsic Argument Strength

Now, we would like to introduce some notions about extrinsic argument strength. As after a deliberation, value information has been added to individual frameworks, firstly, we have it represented in the collective framework. For clarification, we omit the weights on attack relations. Here, we assume the group of agents has a preliminary agreement on value distribution over the common set of arguments.

Definition 3.9 Given $\hat{\mathcal{F}}' = (C_1, \ldots, C_n)$ is a profile of individual m-VAFs where $C_i = (\mathcal{A}_i, \mathcal{R}_i, V_i, \mathcal{M}_i, >_i)$ defined as Definition 2.5, and we assume in any individual m-VAFs $C_i, C_j, \forall A \in \mathcal{A}_i \bigcap \mathcal{A}_j, \mathcal{M}_i^{-1}(A) = \mathcal{M}_j^{-1}(A)$, then one of the possible collective value-based argumentation frameworks w.r.t. $\hat{\mathcal{F}}'$ is $C_m = (\mathcal{A}, \mathcal{R}, V, \mathcal{M}, >_v)$, defined as:

- $\mathcal{A} = \bigcup \mathcal{A}_i; \mathcal{R} = \bigcup \mathcal{R}_i; V = \bigcup V_i;$
- $\mathcal{M} = \mathcal{M}_1 \cup \cdots \cup \mathcal{M}_n$ according to the assumption;

6

- $>_v$ is a strict partial order over V as a deliberated agreement based on the preference profile $(>_1, \ldots, >_n)$.

Note that in individual m-VAFs, $>_i$ could be a strict total or partial order over V_i. Collecting individual preferences requires extra efforts and here we keep them implicit in the process of deliberation. For the reason of that, we simplified $>_v$ as a given information, which can be regarded as the result of deliberation. If we add value information into the running example 2.4, a possible situation for profile $\hat{\mathcal{F}}_1$ could be:

Example 3.10 Given $\hat{\mathcal{F}}_1' = (C_1, C_2, C_3)$ is the profile with value information added to profile $\hat{\mathcal{F}}_1$ in Example 2.4. The value distribution is shown in Figure 2. According to Definition 3.9, the collective value-based argumentation framework m-VAF for $\hat{\mathcal{F}}_1'$ is $C_{m_1} = (\mathcal{A}, \mathcal{R}, V, \mathcal{M}, >_v)$, where $\mathcal{A}$ and $\mathcal{R}$ are unions of the sets of arguments and attack relations in individual frameworks, $V = \{v_1, v_2, v_3, v_4\}$, $\mathcal{M}(v_1) = \{a, d\}, \mathcal{M}(v_2) = \{a, c\}, \mathcal{M}(v_3) = \{b, c\}, \mathcal{M}(v_4) = \{b, d\}$ and a partial order formed after a deliberation as a given information is: $v_1 >_v v_3, v_4 >_v v_2$.

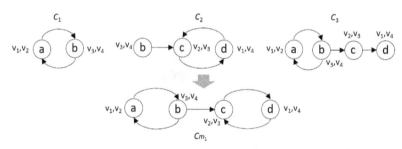

Fig. 2. The profile $\hat{\mathcal{F}}_1'$ and the corresponding C_{m_1} for the running example

As discussed above, we will treat the argument strength acquired from deliberated value preference as the refinement of collective extensions. But given a partial order over values, the extensions of m-VAF can hardly be compared directly. Hence we need to take two steps. First, compute the preference over sets of arguments given value preference and next, compute the ranking of extensions based on the preference over sets of arguments. We adapt some ideas in [20,21] so as to cater for our settings.

Definition 3.11 Let a m-VAF be $C_m = (\mathcal{A}, \mathcal{R}, V, \mathcal{M}, >_v)$, we denote the preference over sets of arguments: strict relation as $\gg$, indifference relation as $\sim$. Given $v_1 >_v v_2$, $\mathcal{M}(v_1) = \{A_1, \ldots, A_n\}$ and $\mathcal{M}(v_2) = \{B_1, \ldots, B_m\}$, we say $\gg$ satisfies $v_1 >_v v_2$ iff $\exists A_i \in \mathcal{M}(v_1)$ s.t. $A_i \gg B_1 \sim \cdots \sim B_m$.

Definition 3.12 We say a set of preferences $>_v$ is consistent only if there exists a model for $\gg$ which satisfies all $v_i >_v v_j$ in $>_v$.

Example 3.13 Continue with Example 3.10, given that $v_1 >_v v_3, v_4 >_v v_2$, according to Definition 3.11, we have:

- the preferences $\gg$ satisfying $v_1 >_v v_3$ are $a \gg b \sim c$ or $d \gg b \sim c$;
- the preferences $\gg$ satisfying $v_4 >_v v_2$ are $b \gg a \sim c$ or $d \gg a \sim c$;

Thus $>_v$ is consistent and the preference $\gg$ satisfying $>_v$ is $d \gg a \sim b \sim c$.

Note that determined by value distribution, it is possible that we can't find a model that satisfies the given value preference, which is called *inconsistency* in Kaci's paper [20]. According to Definition 3.11, $\gg$ is a total pre-order which actually divides sets of arguments in the framework into one or more partitions and constructs the basis for extrinsic argument strength.

Definition 3.14 Given a m-VAF $C_m = (\mathcal{A}, \mathcal{R}, V, \mathcal{M}, >_v)$, and let $\mathcal{F}_C = (\mathcal{A}, \mathcal{R})$ be the abstract argumentation framework without value preference information. Let $E_1, E_2 \in \mathcal{E}_\sigma(\mathcal{F}_C)$, where $\sigma \in \{co, pr, st\}$, if $>_v$ is consistent, then the *extrinsic strength* from E_1 to E_2 with Copland measurement is defined as: $\mathcal{E}_{ex}(E_1, E_2) = |\{A \in E_1 \setminus E_2 : \forall B \in E_2 \setminus E_1 \text{ s.t. } A \gg B\}| - |\{B \in E_2 \setminus E_1 : \forall A \in E_1 \setminus E_2 \text{ s.t. } B \gg A\}|$.

Definition 3.15 Given a m-VAF $C_m = (\mathcal{A}, \mathcal{R}, V, \mathcal{M}, >_v)$ and $\mathcal{F}_C = (\mathcal{A}, \mathcal{R})$, let $E_i, E_j \in \mathcal{E}_\sigma(\mathcal{F}_C)$, where $\sigma \in \{co, pr, st\}$. We also can define the extrinsic-strength matrix, denote as $\mathbf{A}_{ex}$, whose square size equals the cardinality of $\mathcal{E}_\sigma(\mathcal{F}_C)$, with $a_{ij} = \mathcal{E}_{ex}(E_i, E_j)$.

Example 3.16 Continue with 3.13, according to Definition 3.15, the extrinsic-strength matrix $\mathbf{A}_{ex}$ for C_{m_1} is shown in Table 2.

Table 2

$\mathcal{E}_{ex}$	E_1	E_2	E_3
E_1	-	-1	-1
E_2	1	-	0
E_3	1	0	-

Definition 3.17 Given a m-VAF $C_m = (\mathcal{A}, \mathcal{R}, V, \mathcal{M}, >_v)$ and $\mathcal{F}_C = (\mathcal{A}, \mathcal{R})$, the cardinality of $\mathcal{E}_\sigma(\mathcal{F}_C)$ is k. Let $\mathbf{A}_{ex}$ for C_m defined as Definition 3.15, the *extrinsic strength score* for an extension $E_i \in \mathcal{E}_\sigma(\mathcal{F}_C)$, where $\sigma \in \{co, pr, st\}$ is defined as $scor_{ex}(E_i) = \sum_{j=1}^k a_{ij}$. Then the *ranking* on the selection of extensions based on extrinsic argument strength is a total pre-order $\succeq_{ex}$ such that $E_i \succ_{ex} E_j$ iff $scor_{ex}(E_i) > scor_{ex}(E_j)$ and $E_i \sim_{ex} E_j$ iff $scor_{ex}(E_i) = scor_{ex}(E_j)$.

Example 3.18 Continue with Example 3.16, According to Definition 3.17, we have $scor_{ex}(E_1) = -2, scor_{ex}(E_2) = 1, scor_{ex}(E_3) = 1$. Hence the ranking for the extension selection based on extrinsic strength is $E_2 \sim_{ex} E_3 \succ_{ex} E_1$.

3.3 Interplay Mechanisms and Property-based Analysis

In collective argumentation, the source and nature of intrinsic and extrinsic argument strengths are different. The former is based on voting on attack relations collected from individual frameworks, the latter is based on a value ordering and the agreed value distribution on arguments. In our approach, these two different strengths both have influences on the choices for the group, namely a ranking on collective extensions. Although they are both in the form of ranking on extensions, they often guide to different outcomes. It is reasonable, because extrinsic argument strength is disregarding the structure of attack relations and the number of agents which is closely associated with intrinsic argument strength.

Now the question is: how to balance these two argument strengths? Since the result of balance would have direct impact on collective outcomes, what properties (reasons) are provided to support it? In this section, in order to balance these two strengths dialectically, we design α-precedence mechanisms for the interplay. α is a parameter, acting as a balancing weight.

Definition 3.19 The set of all total pre-orders over k alternatives is denoted as $S_{tp}(k)$. We denote $S_{st}(k)$ for the set of all strict total orders over k alternatives.

Definition 3.20 Given $\mathcal{F} = (\mathcal{A}, \mathcal{R})$, let $\mathcal{E}_{\sigma}(\mathcal{F}) = \{E_1, \ldots, E_k\}$ and $E_i, E_j \in \mathcal{E}_{\sigma}(\mathcal{F})$, where $\sigma \in \{co, pr, st\}$. Let $\succcurlyeq_{in}$ and $\succcurlyeq_{ex}$ be the rankings over $\mathcal{E}_{\sigma}(\mathcal{F})$ based on intrinsic and extrinsic argument strength and $x \in \{in, ex\}$. Let $\succsim' \in S_{tp}(k)$ and α is a parameter in $[0, 1]$. The *interplayed ranking* based on α-precedence, denoted as $\succcurlyeq_{\alpha}$, is defined as Equation (1-3) [3] :

$$\Delta(E_i, E_j, \succcurlyeq_x, \succsim') = \begin{cases} 0, & \text{if } E_i \succ_{in} E_j (\text{or } E_i \succ_{ex} E_j) \text{ and } E_i \succ' E_j \\ 0, & \text{if } E_i \sim_{in} E_j (\text{or } E_i \sim_{ex} E_j) \text{ and } E_i \sim' E_j \\ 2, & \text{if } E_i \succ_{in} E_j (\text{or } E_i \succ_{ex} E_j) \text{ and } E_j \succ' E_i \\ 1, & \text{otherwise} \end{cases} \quad (1)$$

$$d(\succcurlyeq_x, \succsim') = \alpha \Sigma_{E_i, E_j \in \mathcal{E}_{\sigma}(\mathcal{F})} \Delta(E_i, E_j, \succcurlyeq_{in}, \succsim') \\ + (1 - \alpha) \Sigma_{E_i, E_j \in \mathcal{E}_{\sigma}(\mathcal{F})} \Delta(E_i, E_j, \succcurlyeq_{ex}, \succsim') \quad (2)$$

$$\succcurlyeq_{\alpha} = \succsim' \quad s.t. \quad argmin \ d(\succcurlyeq_x, \succsim') \quad (3)$$

Definition 3.20 is a Kemeny-style mechanism function [22]. α acts like a weight balancing intrinsic and extrinsic argument strength dialectically. Given a certain value of α, it always selects the ranking over extensions with shortest distance to both strengths.

Definition 3.21 Let α-precedence be defined as Definition 3.20, we further define: if $\alpha = 0.5$, it is called *neutral-precedence*, denote $\alpha_{neutral}$. Let the set of rankings over k extensions yielded by neutral-precede mechanism denoted as $S_{neutral}(k)$.

According to Definition 3.20, when $\alpha = 0$, the interplayed ranking on extensions is actually $\succcurlyeq_{ex}$ and when $\alpha = 1$, the influence from extrinsic argument strength is ignored. Neutral-precedence means the influences from intrinsic and extrinsic argument strength are half-to-half interplayed on the outcomes.

---

[3] *argmin* is argument of the minimum. The simplest example is: argmin f(x) is the value of x for which f(x) attains its minimum.

Proposition 3.22 *Let $\succcurlyeq_{in}$ and $\succcurlyeq_{ex}$ be the rankings over $\mathcal{E}_\sigma(\mathcal{F})$ based on intrinsic and extrinsic argument strengths, then: $\succcurlyeq_{in} \in S_{neutral}(k)$ and $\succcurlyeq_{ex} \in S_{neutral}(k)$.*

Proposition 3.22 indicates that $\succcurlyeq_{in}$ and $\succcurlyeq_{ex}$ are always included in the set of rankings over extensions yielded by neutral-precede mechanism. Besides, $S_{neutral}(k)$ inclines to find a median ranking between the disagreements of $\succcurlyeq_{in}$ and $\succcurlyeq_{ex}$ which is usually in the form of a tie. Now, let us go ahead to define a special α_{well} based on Definition 3.20. The intuition behind it is: it is more well-balanced than above three since it not only has the least bias (nearest to neutral-precedence, we represent it as a penalty term) between two strengths but also yields a strict ranking over extensions without ties.

Definition 3.23 Given $\mathcal{F} = (\mathcal{A}, \mathcal{R})$, let $\mathcal{E}_\sigma(\mathcal{F}) = \{E_1, \ldots, E_k\}$ and $E_i, E_j \in \mathcal{E}_\sigma(\mathcal{F})$, where $\sigma \in \{co, pr, st\}$. Let $\lambda \in \mathbb{R}^+$ be a given penalty parameter [4], $\succ'' \in S_{st}(k)$ and $\alpha \in [0, 1]$. By modifying Equations (1), (2) and (3) in Definition 3.20 to Equations (4), (5) and (6), we define α_{well}-*preceded ranking*, denote $\succcurlyeq_{\alpha_{well}}$ as follows. Let the set of rankings over k extensions yielded by α_{well}-preceded mechanism denoted as $S_{well}(k)$.

$$\Delta'(E_i, E_j, \succcurlyeq_x, \succ'') = \begin{cases} 0, & if \ E_i \succ_{in} E_j (\text{or } E_i \succ_{ex} E_j) \ and \ E_i \succ'' E_j \\ 2, & if \ E_i \succ_{in} E_j (\text{or } E_i \succ_{ex} E_j) \ and \ E_j \succ'' E_i \quad (4) \\ 1, & otherwise \end{cases}$$

$$d'(\succcurlyeq_x, \succ'', \alpha) = \alpha \Sigma_{E_i, E_j \in \mathcal{E}_\sigma(\mathcal{F})} \Delta'(E_i, E_j, \succcurlyeq_{in}, \succ'') + \\ (1 - \alpha) \Sigma_{E_i, E_j \in \mathcal{E}_\sigma(\mathcal{F})} \Delta'(E_i, E_j, \succcurlyeq_{ex}, \succ'') + \lambda \parallel \alpha - 0.5 \parallel \quad (5)$$

Let $\alpha_{well} = \alpha$ and $\succcurlyeq_{\alpha_{well}} = \succ''$ be such that:

$$argmin \ d'(\succcurlyeq_x, \succ'', \alpha) \quad (6)$$

Example 3.24 Continue with Example 3.8 and 3.18. Now we already have $E_3 \succ_{in} E_1 \succ_{in} E_2$ and $E_2 \sim_{ex} E_3 \succ_{ex} E_1$. According to Equation (5), distance functions of six strict total orders over three extensions without the penalty term are shown in Figure 3. We can find that disregarding the penalty term (i.e. $\lambda = 0$), No.2 ordering ($E_3 \succ E_1 \succ E_2$, equals to $\succcurlyeq_{in}$) has the minimal distance and $\alpha_{well} = 1$.

Furthermore, we can observe that as the strength of the penalty term increases, the outputs of the mechanism may vary. As Figure 4 shown, when $\lambda = 1$ and 2, it is No.2 ordering has the minimal distance and $\alpha_{well} = 1$. When $\lambda = 3$, although No.2 ordering still contributes to the minimal distance, the corresponding α_{well} is not a single value but lies in an interval $[0.5, 1]$. When $\lambda > 3$, No.1 ordering ($E_3 \succ E_2 \succ E_1$) joins No.2 ordering as the outputs and $\alpha_{well} = 0.5$.

---

[4] It is also called tuning parameter, controlling the strength of the penalty term. Generally, the influence of penalty term should be positive which means $\lambda > 0$.

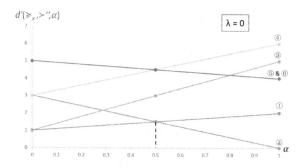

Fig. 3. α_{well} of Example 3.24

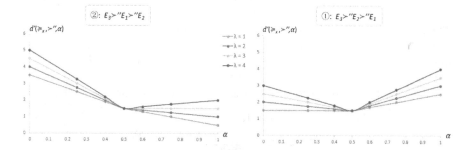

Fig. 4. The relationship between λ and α_{well}, $\succsim_{\alpha_{well}}$

Proposition 3.25 *If $\nexists E_i, E_j \in \mathcal{E}_\sigma(\mathcal{F})$ s.t. $E_i \sim_{in} E_j$ and $E_i \sim_{ex} E_j$, then given a certain penalty parameter $\lambda \in \mathbb{R}^+$, $S_{well}(k) \subseteq S_{neutral}(k)$.*

Proposition 3.25 is the property about the relationship between α_{well}-preceded mechanism and $\alpha_{neutral}$-preceded mechanism. What's more, we are interested in the result of the mechanism, namely the joint influences of two strengths on the collective outcome.

Definition 3.26 Given a $\mathcal{F} = (\mathcal{A}, \mathcal{R})$ and $\mathcal{E}_\sigma(\mathcal{F})$ where $\sigma \in \{co, pr, st, gr\}$. We define the set of *winners* as: $win(\mathcal{E}_\sigma(\mathcal{F}), \succsim_\alpha) = \{E_i | E_i \in \mathcal{E}_\sigma(\mathcal{F})$ and $\nexists E_j \in \mathcal{E}_\sigma(\mathcal{F})$ s.t. $E_j \succsim_\alpha E_i$ but not the case $E_i \succsim_\alpha E_j\}$.

From Definition 3.23 we instantly obtain:

Lemma 3.27 *The α_{well}-preceded rankings over extensions have a unique winner.*

Based on Lemma 3.27, we select some other properties to evaluate our approaches. We adapt P1 proposed in [21] to evaluate preferential argumentative semantics and P2, P3 used for evaluating social rationalities in framework merging [16].

Definition 3.28 Given a $\mathcal{F} = (\mathcal{A}, \mathcal{R})$ and $\mathcal{E}_\sigma(\mathcal{F})$ where $\sigma \in \{co, pr, st, gr\}$. Three properties are defined as follows:

- Property 1 (P1): Extension strictly decrease

11

  · if $|\mathcal{E}_\sigma(\mathcal{F})| > 1$, then $|win(\mathcal{E}_\sigma(\mathcal{F}), \succcurlyeq_\alpha)| < |\mathcal{E}_\sigma(\mathcal{F})|$
- Property 2 (P2): Non-triviality
  · if $|\mathcal{E}_\sigma(\mathcal{F})| > 1$, then $win(\mathcal{E}_\sigma(\mathcal{F}), \succcurlyeq_\alpha) \neq \{\emptyset\}$
- Property 3 (P3): Decisiveness
  · if $|\mathcal{E}_\sigma(\mathcal{F})| > 1$, then $|win(\mathcal{E}_\sigma(\mathcal{F}), \succcurlyeq_\alpha)| = 1$ and $win(\mathcal{E}_\sigma(\mathcal{F}), \succcurlyeq_\alpha) \neq \{\emptyset\}$

P1 concern about the changes in cardinality of the set of extensions compared to the AF without considering intrinsic and extrinsic argument strengths, which implies that the choices for the group may decrease and sceptically accepted arguments may increase after two argument strengths are interplayed. P2 and P3 are related to social rationalities which ensure non-emptiness and uniqueness of collective choices for the group. We adopt above three properties to evaluate our four interplay mechanisms, namely $\succcurlyeq_{in}, \succcurlyeq_{ex}, \succcurlyeq_{\alpha_{neutral}}, \succcurlyeq_{\alpha_{well}}$. The proofs are given in Appendix. The results are presented in Table 3, where ✓ is in any situation (or certain situations) the property is satisfied and × means the property is not always held. From Table 3 we can find that α_{well} is not only well-balanced between intrinsic and extrinsic argument strengths, it satisfies all of three properties (under some certain semantics) as well.

Table 3

mechanisms	P1	P2	P3
$\succcurlyeq_{in}$	×	$\checkmark^{\sigma \in \{pr,st,gr\}}$	×
$\succcurlyeq_{ex}$	×	$\checkmark^{\sigma \in \{pr,st,gr\}}$	×
$\succcurlyeq_{\alpha_{neutral}}$	×	$\checkmark^{\sigma \in \{pr,st,gr\}}$	×
$\succcurlyeq_{\alpha_{well}}$	✓	$\checkmark^{\sigma \in \{pr,st,gr\}}$	$\checkmark^{\sigma \in \{pr,st,gr\}}$

4 Conclusions

Collective argumentation widens the scope for formal argumentation in the area of multi-agent systems. In this paper we identified two possible sources for argument strength and explained the reasons why they act as the refinements of collective extensions. As two argument strengths may lead to conflicted rankings over extensions, we designed α-preceded interplay mechanisms to handle with this problem. Moreover, in order to avoid bias between intrinsic and extrinsic argument strengths and obtain a unique extension as the collective choice, a more specified α_{well}-preceded mechanism was defined. As far as we know, this is the first contribution to the research on different kinds of argument strengths in collective argumentation.

4.1 Related Work

Argument strength has been studied both in structured argumentation [7] and abstract argumentation [17]. In the direction of structured argumentation, argument strength appears in the form of priority orderings over formulae in the language or defeasible inference rules [26,15,24] and some rationality postulates are defined for evaluation [11,18]. In the vein of abstract argumentation, argument strength acts as a preference over arguments as a given information which has abstract nature as the arguments. As it is introduced above, [4,5,21] discussed the roles of preference, Bench-Capon provided a formal basis for the source of preference [6] and Modgil studied the defeasibility of argument strength in abstract argumentation [25]. However, these are works on the basis of a single argumentation framework. When it comes to multiple frameworks, situations become more complicated and more theoretical foundations are needed to deal with the topic of argument strength.

From the perspectives of collective argumentation, argument strength becomes an implicit information in the process of framework merging, closely related with attack relations in individual frameworks and the number of agents. Coste-Marquis adopts a distance-based approach to merge multiple frameworks [14], with argument strength reflected on selecting the majority set of arguments in collective extensions. Delobelle [16], Gabbay [19] and Cayrol [13] adopt a numerical way in framework merging, representing argument strength as weights associated with arguments or attack relations. Unlike Delobelle's approach, the weights in Gabbay's approach can be propogated in the collective framework and the acceptability of arguments is determined by a threshold. Cayrol's approach only defines a quasi-semantics named *vs-defend* to justify a successful defend between pairs of arguments by the weights. In short, both of them are away from standard abstract argumentation semantics (see Definition 2.2). While Delobelle's approach (introduced previously) only identifies one of argument strengths in framework merging, leaving space for our research along this line.

4.2 Future Work

As a preliminary work, there are some open issues for future work. First, we are interested in the selection of winners in collective extensions which promote the values with the highest rank. Second, the sources for argument strength in collective argumentation may more than two kinds and based on different settings. For example, agents may have individual value preferences calling for preference aggregation as an extra operation. Third, if numerical argument strengths are allowed to propagate in the collective framework as it is in Gabbay's work [19], a totally different mechanism of interplay needs to be investigated.

Appendix

Proposition .1 *Let $\succcurlyeq_{in}$ and $\succcurlyeq_{ex}$ be the rankings over $\mathcal{E}_\sigma(\mathcal{F})$ based on intrinsic and extrinsic argument strengths, then: $\succcurlyeq_{in}\in S_{neutral}(k)$ and $\succcurlyeq_{ex}\in S_{neutral}(k)$.*

Proof. Since $\alpha = 0.5$, according to Equations (1), (2), (3) and Definition 3.21, $\succcurlyeq_{\alpha_{neutral}}\in S_{tp}(k)$ is a total pre-order over extensions which minimize the sum of distances to both $\succcurlyeq_{in}$ and $\succcurlyeq_{ex}$ with the same weight 0.5. Actually it is the measure of minimal disagreement between $\succcurlyeq_{in}$ and $\succcurlyeq_{ex}$ in the rankings over extensions. If $\succcurlyeq_{in}=\succcurlyeq_{ex}$, which means there is no disagreement between $\succcurlyeq_{in}, \succcurlyeq_{ex}$, which gives rise to a minimal distance 0 and results in $\succcurlyeq_{\alpha_{neutral}}=\succcurlyeq_{in}=\succcurlyeq_{ex}$.

The conclusion is trivially held. If $\succeq_{in} \neq \succeq_{ex}$, we prove by induction.

- $\succeq_{in}$ disagrees with $\succeq_{ex}$ on the ranking over 1 pair of extensions, for any pair of extensions $E_i, E_j \in \mathcal{E}_\sigma(\mathcal{F})$ there are two possible conditions: (a1) $|\{(E_i, E_j)|E_i \succ_{in} E_j \text{ and } E_j \succ_{ex} E_i\}| = 1$. Then: when $E_i \succ_{\alpha_{neutral}} E_j, E_j \succ_{\alpha_{neutral}} E_i$ or $E_i \sim_{\alpha_{neutral}} E_j$, the distance is minimal, equals to 1. Thus the conclusion is held; (a2) $|\{(E_i, E_j)|E_i \sim_{in} E_j \text{ and } E_j \succ_{ex} E_i\}| = 1$ or $|\{(E_i, E_j)|E_i \succ_{in} E_j \text{ and } E_j \sim_{ex} E_i\}| = 1$. Then: there is no interval between $E_i \succ E_j$ and $E_j \sim E_i$. Thus when $\succeq_{\alpha_{neutral}} = \succeq_{in}$ or $\succeq_{ex}$, the distance is minimal, equals to 0.5.

- Assume $\succeq_{in}$ disagrees with $\succeq_{ex}$ on the ranking over t pair of extensions, the conclusion is held, i.e. $\exists \succeq'_{\alpha_{neutral}} \in S_{neutral}(k)$ which has the minimal distance (suppose to be d_t) equalling to $\succeq_{in}$ and $\succeq_{ex}$. We need to prove the conclusion is also held for $t+1$ pair of extensions. Let $\mathcal{E}_\sigma(\mathcal{F}) = \{E_1, \ldots, E_k\}$, where $\sigma \in \{co, pr, st\}$ and $t+1 \leq k$. For the disagreement on the ranking over the extra 1 pair of extensions, let them be (E_{t_1}, E_{t_2}), there are four possible conditions: (b1) $|\{(E_{t_1}, E_{t_2})|E_{t_1} \succ_{in} E_{t_2} \text{ and } E_{t_2} \succ_{ex} E_{t_1}\}| = 1$ in which the intersection between $\{E_{t_1}, E_{t_2}\}$ and t pairs of arguments is empty. The situation is similar with (a1), i.e. when $E_{t_1} \succ_{\alpha_{neutral}} E_{t_2}, E_{t_2} \succ_{\alpha_{neutral}} E_{t_1}$ or $E_{t_1} \sim_{\alpha_{neutral}} E_{t_2}$, the distance is minimal, equals to $d_t + 1$ and thus the conclusion is held. (b2) $|\{(E_{t_1}, E_{t_2})|E_{t_1} \sim_{in} E_{t_2} \text{ and } E_{t_2} \succ_{ex} E_{t_1}\}| = 1$ or $|\{(E_{t_1}, E_{t_2})|E_{t_1} \succ_{in} E_{t_2} \text{ and } E_{t_2} \sim_{ex} E_{t_1}\}| = 1$ in which the intersection between $\{E_{t_1}, E_{t_2}\}$ and t pairs of arguments is empty. The situation is similar with (a2), i.e. when $\succeq_{\alpha_{neutral}} = \succeq_{in}$ or $\succeq_{ex}$, the distance is minimal, equals to $d_t + 0.5$ and thus the conclusion is held. (b3) The condition of (b1) in which the intersection between $\{E_{t_1}, E_{t_2}\}$ and t pairs of arguments is E_{t_1} or E_{t_2}. It actually means $\succeq_{in}$ disagrees with $\succeq_{ex}$ on the ranking over $t + 2$ pair of extensions. $\succeq_{\alpha_{neutral}} = \succeq_{in}$ or $\succeq_{ex}$, the distance is minimal, equals to $d_t + 2$. (b4) The condition of (b2) in which the intersection between $\{E_{t_1}, E_{t_2}\}$ and t pairs of arguments is E_{t_1} or E_{t_2}. $\succeq_{\alpha_{neutral}} = \succeq_{in}$ or $\succeq_{ex}$, the distance is minimal, equals to $d_t + 2.5$. Therefore, in all of possible conditions, the conclusion is held.

□

Proposition .2 If $\nexists E_i, E_j \in \mathcal{E}_\sigma(\mathcal{F})$ s.t. $E_i \sim_{in} E_j$ and $E_i \sim_{ex} E_j$, then given a certain penalty parameter $\lambda \in \mathbb{R}^+$, $S_{well}(k) \subseteq S_{neutral}(k)$.

Proof. According to Definition 3.23, the penalty term is to minimize the distance between α_{well} and 0.5. Given any profile of $\succeq_{ex}, \succeq_{in} \in S_{tp}(k)$, there exists a value of λ which gives rise to $\alpha_{well} = 0.5$, since the distance according to Equation 5 is a continuous function w.r.t. α in $[0, 1]$. Hence when $\alpha_{well} = 0.5$, according to Definition 3.21, it is $\alpha_{neutral}$. However, $\succeq_{\alpha_{well}} \in S_{st}(k)$, which diverges with $\succeq_{\alpha_{neutral}}$ in yielding only strict orderings over extensions. $\forall E_i, E_j \in \mathcal{E}_\sigma(\mathcal{F})$, only if in the profile, $\succeq_{ex}$ and $\succeq_{in}$ agree on $E_i \sim_x E_j$, where $x \in \{ex, in\}$, $E_i \sim_{\alpha_{neutral}} E_j$ but not the case for $\succeq_{\alpha_{well}}$. The rest of possible situations, no matter $\succeq_{ex}$ and $\succeq_{in}$ agree on a strict ordering or disagree on a strict ordering over E_i, E_j (one of them might be a tie), according

14

to Proposition 3.22, there exists a strict ordering in $S_{neutral}(k)$ which shares the same minimal distance with other members in $S_{neutral}(k)$ which is also in $S_{well}(k)$. Therefore, the conclusion is held. □

Proposition .3 $\succcurlyeq_{in}, \succcurlyeq_{ex}$ and $\succcurlyeq_{\alpha_{neutral}}$ don't satisfy Property 1, while $\succcurlyeq_{\alpha_{well}}$ satisfies it.

Proof. Given $\mathcal{F} = (\mathcal{A}, \mathcal{R})$, when $|\mathcal{E}_\sigma(\mathcal{F})| > 1$, $\mathcal{E}_\sigma(\mathcal{F})$ is not possible to be the set of grounded extensions which has the cardinality of 1. Therefore, $\sigma = gr$ is always held for Property 1. As we know, $\succcurlyeq_{in}, \succcurlyeq_{ex}, \succcurlyeq_{\alpha_{neutral}}$ are all total pre-order over extensions. Thus when $\sigma \in \{co, pr, st\}$, let the cardinality of $\mathcal{E}_\sigma(\mathcal{F})$ be k. As a total pre-order over k extensions, it is possible that $win(\mathcal{E}_\sigma(\mathcal{F}), \succcurlyeq_{in}) = \mathcal{E}_\sigma(\mathcal{F})$ which leads to $|win(\mathcal{E}_\sigma(\mathcal{F}), \succcurlyeq_{in})| = |\mathcal{E}_\sigma(\mathcal{F})|$. Thus $\succcurlyeq_{in}$ doesn't satisfy Property 1, so do $\succcurlyeq_{ex}, \succcurlyeq_{\alpha_{neutral}}$. While According to Lemma 3.27, $\succcurlyeq_{\alpha_{well}}$ yield a linear order over k extensions, which means $|win(\mathcal{E}_\sigma(\mathcal{F}), \succcurlyeq_{\alpha_{well}})| = 1$. Therefore, given $|\mathcal{E}_\sigma(\mathcal{F})| > 1$, $|win(\mathcal{E}_\sigma(\mathcal{F}), \succcurlyeq_{\alpha_{well}})| < |\mathcal{E}_\sigma(\mathcal{F})|$. □

Proposition .4 Only if $\sigma = \{pr, st, gr\}$, $\succcurlyeq_{in}, \succcurlyeq_{ex}, \succcurlyeq_{\alpha_{neutral}}$ and $\succcurlyeq_{\alpha_{well}}$ satisfy Property 2.

Proof. Given $\mathcal{F} = (\mathcal{A}, \mathcal{R})$ and $|\mathcal{E}_\sigma(\mathcal{F})| > 1$, $\mathcal{E}_\sigma(\mathcal{F})$ is not possible to be the set of grounded extensions which has the cardinality of 1. Therefore, $\sigma = gr$ is always held for Property 1. Given $\sigma \in \{pr, st\}$, assume that $win(\mathcal{E}_\sigma(\mathcal{F}), \succcurlyeq_\alpha) = \{\emptyset\}$, which means $\emptyset \in \mathcal{E}_{pr}(\mathcal{F})$ or $\emptyset \in \mathcal{E}_{st}(\mathcal{F})$, and except $\emptyset$, there are non-empty preferred or stabled extensions. It is impossible for preferred semantics, because it is a maximal(w.r.t. set inclusion) complete extension of $\mathcal{F}$ and $\emptyset$ is included in any non-empty extension. It is also a contradiction for stable semantics, because according to the definition, $\emptyset$ can't attack all the arguments which do not belong to it. Thus if $\sigma \in \{pr, st\}$ and $|\mathcal{E}_\sigma(\mathcal{F})| > 1$, $win(\mathcal{E}_\sigma(\mathcal{F}), \succcurlyeq_\alpha) \neq \{\emptyset\}$. We could find out that when $\sigma = co$, Property 2 doesn't hold for these mechanisms, for the reason that according to the definition 2.2, complete extension contains the grounded extension and preferred extensions (w.r.t set inclusion) and the grounded extension is possible to be empty while preferred extensions are not empty. Therefore, even if $|\mathcal{E}_{co}(\mathcal{F})| > 1$, $win(\mathcal{E}_{co}(\mathcal{F}), \succcurlyeq_\alpha)$ is possible to be empty. □

Proposition .5 $\succcurlyeq_{in}, \succcurlyeq_{ex},$ and $\succcurlyeq_{\alpha_{neutral}}$ don't satisfy Property 3, while $\succcurlyeq_{\alpha_{well}}$ satisfies it only if $\sigma \in \{pr, st, gr\}$.

Proof. Given $\mathcal{F} = (\mathcal{A}, \mathcal{R})$ and $|\mathcal{E}_\sigma(\mathcal{F})| > 1$, $\mathcal{E}_\sigma(\mathcal{F})$ is not possible to be the set of grounded extensions which has the cardinality of 1. Therefore, $\sigma = gr$ is always held for Property 3. Since $|\mathcal{E}_\sigma(\mathcal{F})| > 1$, the proof for $win(\mathcal{E}_\sigma(\mathcal{F}), \succcurlyeq_\alpha) \neq \{\emptyset\}$ is same as Proposition?? as it holds for $\succcurlyeq_{in}, \succcurlyeq_{ex}, \succcurlyeq_{\alpha_{neutral}}$ and $\succcurlyeq_{\alpha_{well}}$ only if $\sigma \in \{pr, st, gr\}$. As $\succcurlyeq_{in}, \succcurlyeq_{ex}, \succcurlyeq_{\alpha_{neutral}}$ are total pre-orders over extensions, $|win(\mathcal{E}_\sigma(\mathcal{F}), \succcurlyeq_\alpha)| = 1$ doesn't always hold for them. While $\succcurlyeq_{\alpha_{well}}$ yields a strict total order over extensions, therefore, when $\mathcal{E}_\sigma(\mathcal{F}) \neq \emptyset$, α_{well}-preceded ranking can select the unique winner for $\mathcal{E}_\sigma(\mathcal{F})$, i.e. $|win(\mathcal{E}_\sigma(\mathcal{F}), \succcurlyeq_{\alpha_{well}})| = 1$. □

References

[1] Amgoud, L. and C. Cayrol, *Inferring from inconsistency in preference-based argumentation frameworks*, Journal of Automated Reasoning **29** (2002), pp. 125–169.

[2] Amgoud, L. and C. Cayrol, *On the acceptability of arguments in preference-based argumentation*, arXiv preprint arXiv:1301.7358 (2013).

[3] Amgoud, L., S. Parsons and N. Maudet, *Arguments, dialogue, and negotiation*, a a **10** (2000), p. 02.

[4] Amgoud, L. and S. Vesic, *Two roles of preferences in argumentation frameworks*, in: *European Conference on Symbolic and Quantitative Approaches to Reasoning and Uncertainty*, Springer, 2011, pp. 86–97.

[5] Amgoud, L. and S. Vesic, *Rich preference-based argumentation frameworks*, International Journal of Approximate Reasoning **55** (2014), pp. 585–606.

[6] Bench-Capon, T. J., *Persuasion in practical argument using value-based argumentation frameworks*, Journal of Logic and Computation **13** (2003), pp. 429–448.

[7] Besnard, P., A. Garcia, A. Hunter, S. Modgil, H. Prakken, G. Simari and F. Toni, *Introduction to structured argumentation*, Argument & Computation **5** (2014), pp. 1–4.

[8] Bodanza, G., F. Tohmé and M. Auday, *Collective argumentation: A survey of aggregation issues around argumentation frameworks*, Argument & Computation **8** (2017), pp. 1–34.

[9] Booth, R., *Judgment aggregation in abstract dialectical frameworks*, in: *Advances in Knowledge Representation, Logic Programming, and Abstract Argumentation*, Springer, 2015 pp. 296–308.

[10] Booth, R., E. Awad and I. Rahwan, *Interval methods for judgment aggregation in argumentation*, in: *Fourteenth International Conference on Principles of Knowledge Representation and Reasoning (KR-14)*, AAAI Press, 2014, pp. 594–597.

[11] Caminada, M. and L. Amgoud, *On the evaluation of argumentation formalisms*, Artificial Intelligence **171** (2007), pp. 286–310.

[12] Caminada, M. and G. Pigozzi, *On judgment aggregation in abstract argumentation*, Autonomous Agents and Multi-Agent Systems **22** (2011), pp. 64–102.

[13] Cayrol, C. and M.-C. Lagasquie-Schiex, *Weighted argumentation systems: A tool for merging argumentation systems*, in: *2011 IEEE 23rd International Conference on Tools with Artificial Intelligence*, IEEE, 2011, pp. 629–632.

[14] Coste-Marquis, S., C. Devred, S. Konieczny, M.-C. Lagasquie-Schiex and P. Marquis, *On the merging of dung's argumentation systems*, Artificial Intelligence **171** (2007), pp. 730–753.

[15] Čyras, K. and F. Toni, *Aba+: assumption-based argumentation with preferences*, arXiv preprint arXiv:1610.03024 (2016).

[16] Delobelle, J., S. Konieczny and S. Vesic, *On the aggregation of argumentation frameworks: operators and postulates*, Journal of Logic and Computation **28** (2018), pp. 1671–1699.

[17] Dung, P. M., *On the acceptability of arguments and its fundamental role in nonmonotonic reasoning, logic programming and n-person games*, Artificial intelligence **77** (1995), pp. 321–357.

[18] Dung, P. M. and P. M. Thang, *Fundamental properties of attack relations in structured argumentation with priorities*, Artificial Intelligence **255** (2018), pp. 1–42.

[19] Gabbay, D. and O. Rodrigues, *A numerical approach to the merging of argumentation networks*, in: *International Workshop on Computational Logic in Multi-Agent Systems*, Springer, 2012, pp. 195–212.

[20] Kaci, S. and L. van der Torre, *Preference-based argumentation: Arguments supporting multiple values*, International Journal of Approximate Reasoning **48** (2008), pp. 730–751.

[21] Kaci, S., L. van der Torre and S. Villata, *Preference in abstract argumentation*, , **305**, IOS Press, 2018, pp. 405–412.

[22] Kemeny, J. G., *Mathematics without numbers*, Daedalus **88** (1959), pp. 577–591.

[23] Kok, E. M., J.-J. C. Meyer, H. Prakken and G. A. Vreeswijk, *A formal argumentation framework for deliberation dialogues*, in: *International Workshop on Argumentation in Multi-Agent Systems*, Springer, 2010, pp. 31–48.

[24] Liao, B., N. Oren, L. van der Torre and S. Villata, *Prioritized norms in formal argumentation*, Journal of Logic and Computation **29** (2019), pp. 215–240.

[25] Modgil, S., *Reasoning about preferences in argumentation frameworks*, Artificial intelligence **173** (2009), pp. 901–934.

[26] Modgil, S. and H. Prakken, *A general account of argumentation with preferences*, Artificial Intelligence **195** (2013), pp. 361–397.

[27] Prakken, H., *Coherence and flexibility in dialogue games for argumentation*, Journal of logic and computation **15** (2005), pp. 1009–1040.

[28] Rienstra, T., M. Thimm and N. Oren, *Opponent models with uncertainty for strategic argumentation*, in: *Twenty-Third International Joint Conference on Artificial Intelligence*, Citeseer, 2013.

[29] Roth, B., R. Riveret, A. Rotolo and G. Governatori, *Strategic argumentation: a game theoretical investigation*, in: *Proceedings of the 11th international conference on artificial intelligence and law*, 2007, pp. 81–90.

[30] Thimm, M., *Strategic argumentation in multi-agent systems*, KI-Künstliche Intelligenz **28** (2014), pp. 159–168.

A Flexible Approach to Argumentation Framework Analysis using Theorem Proving

David Fuenmayor [1]   Alexander Steen [1]

University of Luxembourg, Department of Computer Science
6, avenue de la Fonte
L-4364 Esch-sur-Alzette, Luxembourg

Abstract

Argumentation frameworks constitute the central concept of argumentation theory of Dung. In this paper we present a novel and flexible approach of analyzing argumentation frameworks and their semantics based on an encoding into extensional type theory (classical higher-order logic). This representation enables the use of a wide range of interactive and automated higher-order reasoning tools for assessing argumentation frameworks. This includes the generation of labellings (and extensions), the assessment of meta-theoretic properties, the conduction of interactive empirical experiments, and the flexible analysis of argumentation frameworks with interpreted arguments.

Keywords: Abstract Argumentation, Extensional Type Theory, Automated Reasoning, Meta-logical reasoning, Proof Assistants.

1   Introduction

Argumentation theory is an active field of research in AI. Argumentation frameworks [12] constitute the central concept in abstract argumentation, they have many topical applications in, among others, logic programming, multi-agent systems, non-monotonic reasoning. Since the original formulation of Dung, a lot of research was conducted concerning algorithmic procedures, complexity aspects, as well as various extended and related formalisms (cf. [3] and references therein; cf. also [2] for a comprehensive survey).

In this paper, we propose to investigate argumentation frameworks from the perspective of extensional type theory (ExTT), also commonly simply referred to as *higher-order logic*. To that end, we present an encoding [2] of ar-

---

[1] E-Mail: {david.fuenmayor, alexander.steen}@uni.lu. Authors are sorted alphabetically by last name. Both authors acknowledge financial support from the Luxembourg National Research Fund (FNR) under grant CORE AuReLeE (C20/IS/14616644).

[2] This is analogous to a *shallow semantical embedding* [4]. We encode the semantics of an *object logic* (in this case: formal argumentation notions) as syntactic abbreviations involving ExTT expressions. *Deep embeddings*, by contrast, explicitly introduce recursive structures representing object-logical formulas together with inductively defined predicates (e.g. *eval*).

gumentation frameworks and their semantics into higher-order logic and make use of available reasoning tools from the (higher-order) automated reasoning community. In particular, we demonstrate how this setup can be used as a framework to (a) flexibly synthesize labellings for argumentation frameworks that satisfy arbitrarily complex properties, (b) conduct (explorative) analyses of meta-logical properties, and (c) formalize rich instantiations of argumentation frameworks in which the arguments can be formulas of some expressive (non-classical) logic.

The experiments presented in this paper were conducted using the proof assistant Isabelle/HOL [17] (cf. §2.2 for some details). The corresponding Isabelle/HOL source files (so-called *theory files*) for this work are freely available at GitHub [14].

The layout of the paper is as follows: In §2 we briefly introduce the concepts of argumentation frameworks and some introductory information on ExTT. §3 presents the encoding of argumentation frameworks and their semantics into higher-order logic. Subsequently, an emphasis is put on the generation of labellings in §4, and an outlook of further application perspectives in given in §5. Finally, we briefly conclude in §6.

2   Preliminaries

The notion of argumentation frameworks and their semantics are introduced. Also, a brief exposition to extensional type theory is given. In the remainder of this paper, the latter formalism will be used for modeling the former.

2.1   Abstract Argumentation

In abstract argumentation theory of Dung [12], arguments are represented as abstract objects and constitute the nodes of a directed graph. The edges of this graph represent (directed) attacks between arguments. This is formalized by the well-known structure of argumentation frameworks. [3]

Definition 2.1 An *argumentation framework* AF is a pair $AF = (A, \rightarrow)$, where A is a finite set and $\rightarrow \subseteq A \times A$ is a binary relation on A. The elements of A are called *arguments*, and $\rightarrow$ is also referred to as the *attack relation*.

An argumentation framework essentially gives an overview of relevant arguments and how they interact (e.g., conflict). Given an argumentation framework AF, one of the main tasks is to determine the subsets of arguments that can be reasonably accepted and those that have to be rejected. This is addressed by so-called argumentation semantics that impose certain restrictions on this selection. There are two standard approaches for argumentation semantics: The more traditional extension-based semantics [12] and the popular labelling-based semantics [1]:

---

[3] This brief introduction largely follows the survey of Baroni, Caminada and Giacomin [1], to which we refer to for further details on argumentation frameworks and their semantics.

Definition 2.2 An *extension-based semantics* S associates with each argumentation framework $AF = (A, \rightarrow)$ a set of *extensions*, denoted $\mathcal{E}_S(AF)$, where $\mathcal{E}_S(AF) \subseteq 2^A$.

Roughly speaking, an extension is the subset of all arguments that are accepted (under some criterion given by S), while the others are rejected.

Definition 2.3 Let $AF = (A, \rightarrow)$ be an argumentation framework. A *labelling* of AF is a function $\mathcal{L}ab : A \Rightarrow \{\text{In}, \text{Out}, \text{Undec}\}$, the set of all labellings of AF is denoted $\mathfrak{L}(AF)$. A *labelling-based semantics* S then associates with each AF a set of *labellings*, denoted $\mathcal{L}_S(AF)$, where $\mathcal{L}_S(AF) \subseteq \mathfrak{L}(AF)$.

Intuitively, the labels In and Out represent the status of accepting and rejecting a given argument, respectively. Arguments labelled Undec are left undecided, either because one explicitly refrains from accepting resp. rejecting it, or because it cannot be labelled otherwise.

The classical (extension-based) argumentation semantics of Dung are called *conflict-free, admissible, complete, grounded, preferred* and *stable* [12]. Furthers include, e.g., *ideal* semantics [13] and *semi-stable* semantics [9]. For each of these semantics there exists an equivalent labelling-based formulation, and each extension can be translated into a labelling and vice versa [10,1].

We omit the formal definitions of the semantics at this point, as they will be subject of the discussions in §3.

2.2   Extensional Type Theory

Extensional type theory (ExTT) is an expressive higher-order logical formalism based on a simply typed λ-calculus which originates from works of Church, Henkin and others [11,15].

The term *higher-order* refers to the ability of ExTT to represent quantifications over predicate and function variables — as opposed to first-order logics, in which quantification is restricted to individuals only. Furthermore ExTT provides λ-notation as an expressive binding mechanism to denote unnamed functions, predicates and sets (by their characteristic functions), and it comes with built-in principles of Boolean and functional extensionality as well as type-restricted comprehension. ExTT constitutes the foundation of most contemporary higher-order automated reasoning systems [6].

Reasoning Systems. Interactive theorem provers (also referred to as *proof assistants*) are systems that allow for the creation and assessment of computer-verified formal proofs, and also facilitate interactive experiments on given formal representations. They are usually based on (extensions of) higher-order logic; one well-established example is Isabelle/HOL [17] that is employed in a wide range of applications, including this paper.

One of the most relevant practical features of Isabelle/HOL is the Sledgehammer system [7] that bridges between the proof assistant and external automated theorem proving (ATP) systems, such as the first-order ATP system E [18] or the higher-order ATP system Leo-III [19]. The idea is to use these automated systems to autonomously solve proof obligations and to import the proofs into the verified context of Isabelle/HOL. The employment of Sledge-hammer is of great practical importance and usually a large amount of laborious proof engineering work can be solved by the ATP systems. [4]

3   Encoding of Argumentation Semantics

In this section we present the encoding of argumentation semantics in ExTT, using a syntactical representation close to the one of Isabelle/HOL.

A few technical remarks are in order: In Isabelle/HOL types are either base types, type variables or functional types (the type of functions). In the remainder, o denotes the type of Booleans (i.e., formulas), and $'a$ is a type variable representing an arbitrary type. Function types are denoted $\tau \Rightarrow \nu$, where τ and ν are themselves types. The usual classical connectives are given by $\neg$, $\vee$, $\wedge$, $\longrightarrow$ and $\longleftrightarrow$ for negation, disjunction, conjunction, implication and equivalence, respectively. Universally and existentially quantified formulas are denoted by $\forall X.\, s$ and $\exists X.\, s$, respectively, and anonymous functions are written $\lambda X.\, s$ (where X is an arbitrary identifier and s is a formula).

Further interpreted syntactical notions can be introduced using Is-abelle/HOL's meta-logical theory file syntax: A *definition* defines a new symbol that can be regarded (for the purposes of this paper) an abbreviation for terms; we will write $c := s$ to denote the introduction of a new symbol c with defini-tion s, where s is some term, in the following. A *type synonym* is similar to a term definition but rather introduces a new type symbol that abbreviated a (complex) type expression.

Isabelle/HOL formalizes proofs using the general purpose proof language *Isar* [20] which is part of the meta-logical language level of the system. Such formal and internally verified proofs might also be generated by Sledgeham-mer using external ATP systems. Additionally, counter-models finders such as *Nitpick* [8] may be used to refute given conjectures by providing, if successful, specific counterexamples.

3.1   Basic Notions on Sets and Orderings

The definitions and results discussed below are found in [14, theory `base.thy`].

We start by defining useful type synonyms for the types of sets and relations, which will be represented by characteristic functions (i.e., predicates) on objects of some type $'a$. We thus define $'a$ `Set` and $'a$ `Rel` as type synonyms for the function types $'a \Rightarrow o$ and $'a \Rightarrow 'a \Rightarrow o$, respectively. Set equality and the subset relation can be defined as terms of type $'a$ `Set` $\Rightarrow$ $'a$ `Set` $\Rightarrow o$ as follows (all set operations are written as infix operators in the remainder):

$$A \approx B \;:=\; \forall x.\, (A\,x) \longleftrightarrow (B\,x) \qquad A \subseteq B \;:=\; \forall x.\, (A\,x) \longrightarrow (B\,x)$$

Analogously, the remaining set-theoretic operations can be defined by anony-

---

[4] In fact, all of the formalized proofs within Isabelle/HOL presented in the remainder of this paper were generated automatically using Sledgehammer.

mous functions reducing it to the respective underlying logical connectives:

$$A \cap B := \lambda w.\ (A\ w) \wedge (B\ w) \qquad A \cup B := \lambda w.\ (A\ w) \vee (B\ w)$$
$$-A := \lambda w.\ \neg(A\ w)$$

where $\cap$ and $\cup$ are both terms of type $'a\ \mathtt{Set} \Rightarrow\ 'a\ \mathtt{Set} \Rightarrow\ 'a\ \mathtt{Set}$ and $-$ is of type $'a\ \mathtt{Set} \Rightarrow\ 'a\ \mathtt{Set}$.

Because of their importance in various argumentation semantics, we additionally provide generic notions for representing minimal and maximal (resp. least and greatest) sets, with respect to set inclusion: Let Obj be a term of some type τ, and $Prop$ a predicate of type $\tau \Rightarrow o$. We formalize the statement that the set $S(Obj)$ induced by Obj is minimal/maximal/least/greatest among all objects O satisfying property $Prop$ as follows:

$$
\begin{aligned}
minimal\ Prop\ Obj\ S\ &:=\ Prop\ Obj\ \wedge \\
&\quad (\forall O.\ Prop\ O\ \wedge\ S(O) \subseteq S(Obj) \longrightarrow S(O) \approx S(Obj)) \\
maximal\ Prop\ Obj\ S\ &:=\ Prop\ Obj\ \wedge \\
&\quad (\forall O.\ Prop\ O\ \wedge\ S(Obj) \subseteq S(O) \longrightarrow S(O) \approx S(Obj)) \\
least\ Prop\ Obj\ S\ &:=\ Prop\ Obj\ \wedge \\
&\quad (\forall O.\ Prop\ O\ \longrightarrow\ S(Obj) \subseteq S(O)) \\
greatest\ Prop\ Obj\ S\ &:=\ Prop\ Obj\ \wedge \\
&\quad (\forall O.\ Prop\ O\ \longrightarrow\ S(O) \subseteq S(Obj))
\end{aligned}
$$

In fact, we formally verified in Isabelle/HOL that, based on these definitions, a least (resp. greatest) set is minimal (resp. maximal) while obtaining counter-models for the converse (using Nitpick). Also, it can be shown that least and greatest elements are unique and that minimal/maximal elements collapse to the least and greatest one when the latter exist. We have also verified some useful results concerning the existence of least/greatest fixed points of monotone functions.

3.2 Extension-based semantics

The definitions and results discussed below are found in [14, theory `extensions.thy`].

In ExTT, an argumentation framework AF is completely characterized by its underlying attack relation $\rightarrow$ of type $'a\ \mathtt{Rel}$, since the set of arguments (i.e., the carrier of $\rightarrow$) is given implicitly as the set of objects of type $'a$. We thus use the type synonym $'a\ \mathcal{AF}$ for argumentation frames as shorthand for $'a\ \mathtt{Rel}$; equaling to type of the underlying attack relation. We will implicitly assume in the following that, in the context of an argumentation framework AF of type $'a\ \mathcal{AF}$, a set of arguments is a term of type $'a\ \mathtt{Set}$.

Given an argumentation framework AF and a set of arguments S, we define the set of attacked and attacking arguments, denoted $[AF|S]^{+}$ and $[AF|S]^{-}$, respectively, as follows:

$$[AF|S]^{+} := \{b\,|\,\exists a.\ S\ a\ \wedge\ AF\ a\ b\} \qquad [AF|S]^{-} := \{b\,|\,\exists a.\ S\ a\ \wedge\ AF\ b\ a\}.$$

We can now define the fundamental notion of *defense* (aka. *acceptability* in [12]) of arguments:

Definition 3.1 Let A be an argument and S a set of arguments. We say that S *defends* A if each argument B attacking A is itself attacked by at least one argument in S.

This is encoded as a term of type $`a\ \mathcal{AF} \Rightarrow\ `a\ \mathsf{Set} \Rightarrow\ `a \Rightarrow o$ as follows:

$$\textit{defends } AF\ S\ a\ :=\ \forall b.\ AF\ b\ a\ \longrightarrow (\exists z.\ S\ z \wedge AF\ z\ b)$$

In fact, Isabelle's simplifier can verify automatically that this definition corresponds to $[AF|\{a\}]^- \subseteq [AF|S]^+$.

The notion of a *characteristic function* $\mathcal{F}$ of an argumentation framework AF can in fact simply be defined as an alias for the function *defends*, which gives us:

$$\mathcal{F}\ AF\ S\ =\ \lambda a.\ \textit{defends } AF\ S\ a$$

It can easily be verified in Isabelle/HOL that $\mathcal{F}$ (i.e. *defends*) is indeed a monotone function and that it has both a least and greatest fixed point.

The well-known extension-based semantics of Dung [12] have been encoded based on the definitions above (cf. [14, theory `extensions.thy`] for details). We leave them out here, due to space limitations, and focus only on the labelling-based semantics in the following; however, all results in the remainder are equally applicable to the extension-based semantics.

3.3   Labelling-based semantics

The contents of this section are contained in the theories `labellings.thy` and `tests.thy` in the corresponding Isabelle sources [14].

Note that we have encoded sets (i.e., potential argument *extensions*) as functions mapping objects of an arbitrary type $`a$ (i.e., arguments) to the two-element Boolean type o. Generalising on this, we can now define *labellings* as functions into some arbitrary but finite codomain of *labels*. We follow the usual approach given in Def. 2.3 and assume the a set of three labels $\{\mathsf{In}, \mathsf{Out}, \mathsf{Undec}\}$, by defining the Isabelle/HOL datatype: [5]

$$\mathsf{Label} := \mathsf{In} \mid \mathsf{Out} \mid \mathsf{Undec},$$

together with $`a\ \mathsf{Labelling}$ as type synonym for the type $`a \Rightarrow \mathsf{Label}$.

Definition 3.2 Given a labelling $\mathcal{L}ab$ we write $in(\mathcal{L}ab)$, $out(\mathcal{L}ab)$, and $undec(\mathcal{L}ab)$ (read as *in-set*, *out-set*, *undec-set*, respectively) for the sets of arguments labelled by $\mathcal{L}ab$ as In, Out and Undec, respectively.

We encode this definition in Isabelle/HOL as:

$$in(\mathcal{L}ab) := \lambda x.\ \mathcal{L}ab(x) = \mathsf{In}$$
$$out(\mathcal{L}ab) := \lambda x.\ \mathcal{L}ab(x) = \mathsf{Out}$$
$$undec(\mathcal{L}ab) := \lambda x.\ \mathcal{L}ab(x) = \mathsf{Undec}$$

---

[5] A datatype can be, in turn, encoded into plain ExTT.

Now that we have provided means to represent the *as-is* state of an argument wrt. a given labelling, we can additionally represent a target situation in which an argument is said to be adequately or *legally* labelled.

Definition 3.3 Let a be a argument. a is said to be *legally in* if all of its attackers are labelled Out. a is said to be *legally out* if it has at least one attacker that is labelled In. a is said to be *legally undecided* if it is neither *legally in* nor *legally out*.

In Isabelle/HOL, we encode the above notions by means of predicates of type $`a\ \mathcal{AF} \Rightarrow\ `a$ Labelling $\Rightarrow\ `a \Rightarrow o$ as follows:

$$legallyIn\ AF\ Lab := \lambda a.\forall b.(AF\ b\ a) \longrightarrow out\ Lab\ b$$
$$legallyOut\ AF\ Lab := \lambda a.\exists b.(AF\ b\ a) \longrightarrow in\ Lab\ b$$
$$legallyUndec\ AF\ Lab := \lambda a.\neg(legallyIn\ AF\ Lab\ a) \wedge \neg(legallyOut\ AF\ Lab\ a)$$

Finally, employing the definitions above, the well-known notions of conflict-free and admissible labellings can be defined.

Definition 3.4 [Conflict-free labelling; cf. [1, Def. 16]] A labelling $\mathcal{L}ab$ is termed *conflict-free* if (i) every In-labelled argument is not *legally out*; and (ii) every Out-labelled argument is *legally out*.

Definition 3.5 [Admissible labelling; cf. [1, Def. 10]] A labelling $\mathcal{L}ab$ is termed *admissible* if (i) every In-labelled argument is *legally in*; and (ii) every Out-labelled argument is *legally out*.

The two definitions above have been encoded in Isabelle/HOL as predicates of type $`a\ \mathcal{AF} \Rightarrow\ `a$ Labelling $\Rightarrow o$ as follows:

$$conflictfree\ AF\ Lab := \forall x.(in\ Lab) \longrightarrow \neg legallyOut\ AF\ Lab) \wedge$$
$$(out\ Lab) \longrightarrow legallyOut\ AF\ Lab$$
$$admissible\ AF\ Lab := \forall x.(in\ Lab) \longrightarrow legallyIn\ AF\ Lab) \wedge$$
$$(out\ Lab) \longrightarrow legallyOut\ AF\ Lab$$

We can, in fact, employ Isabelle to verify automatically that admissible labellings always exist (e.g., consider a labelling where all arguments are Undec) and also that admissible labellings are indeed conflict-free.

Moreover, it can be proven automatically that, for admissible labellings, if an argument is *legally undec* then it is labelled Undec, but not the other way round (counter-models provided by Nitpick). Interestingly, one can also verify, again by generating counter-models with Nitpick, that for admissible labellings, a *legally in* (resp. *legally out*) argument is not generally labelled In (resp. Out). This situation changes, however, when we start considering complete labellings.

Definition 3.6 [Complete labelling; cf. [1, Def. 18]] A labelling $\mathcal{L}ab$ is termed *complete* if (i) it is admissible; and (ii) every Undec-labelled argument is *legally undec*.

The corresponding Isabelle/HOL encoding is given by:

$$complete\ AF\ Lab := admissible\ AF\ Lab\ \wedge$$
$$(\forall x.\ undec\ Lab\ x \longrightarrow legallyUndec\ AF\ Lab\ x)$$

Using the Sledgehammer tool from within of Isabelle/HOL it can be proven automatically that for complete labellings, *legally in* (resp. *legally out*) arguments are indeed labelled In (resp. Out). In fact, this alternative definition for a complete labelling has been verified as a theorem:

$$complete\ AF\ Lab = \forall x.(in\ Lab\ x \longleftrightarrow legallyIn\ AF\ Lab\ x)\ \wedge$$
$$(out\ Lab\ x \longleftrightarrow legallyOut\ AF\ Lab\ x)$$

Also, we can verify automatically that every complete labelling is admissible but not the other way round (since counter-models are found by Nitpick).

In fact, we can prove that for complete labellings, we have that *in/out-sets* completely determine the labelling, i.e., it holds that

$$(complete\ AF\ L1 \wedge complete\ AF\ L2) \longrightarrow \big(in\ L1 \approx in\ L2 \longrightarrow (L1 = L2)\big)$$

and

$$(complete\ AF\ L1 \wedge complete\ AF\ L2) \longrightarrow \big(out\ L1 \approx out\ L2 \longrightarrow (L1 = L2)\big).$$

By generating counterexamples with Nitpick we verified that, in contrast, *undec-sets* do not completely determine the (complete) labellings. Another automatically verified result is the following:

$$(complete\ AF\ L1 \wedge complete\ AF\ L2) \longrightarrow (in\ L1 \subseteq in\ L2 \longleftrightarrow out\ L1 \subseteq out\ L2)$$

We now turn to the notions of minimality and maximality for complete labellings, drawing upon the definitions provided in §3.1. We have verified several properties and interrelations for them. As an example, we have shown that given a complete labelling $\mathcal{L}ab$, minimality of in($\mathcal{L}ab$) is equivalent to minimality of out($\mathcal{L}ab$), i.e., it holds that

$$minimal(complete\ AF)\ Lab\ in = minimal(complete\ AF)\ Lab\ out.$$

With the results above we are now in a position to discuss labellings where *in-sets* are minimal or maximal. They correspond to the so-called *grounded* resp. *preferred* labellings.

Definition 3.7 [Grounded labelling; cf. [1, Def. 20]] A labelling $\mathcal{L}ab$ is termed *grounded* if it is a (in fact: *the*) complete labelling whose *in-set* is minimal (wrt. set inclusion) among all the complete labellings.

Definition 3.8 [Preferred labelling; cf. [1, Def. 22]] A labelling $\mathcal{L}ab$ is termed *preferred* if it is a complete labelling whose *in-set* is maximal (wrt. set inclusion) among all the complete labellings.

The two definitions above are encoded in Isabelle/HOL as follows:

$$grounded\ AF\ Lab := minimal\ (complete\ AF)\ Lab\ in$$
$$preferred\ AF\ Lab := maximal\ (complete\ AF)\ Lab\ in$$

which, recalling the definitions of minimality/maximality in §3.1 unfolds into

$$grounded\ AF\ Lab := complete\ AF\ Lab\ \wedge$$
$$(\forall L.complete\ AF\ L \wedge in(L) \subseteq in(Lab) \longrightarrow in(L) \approx in(Lab))$$
$$preferred\ AF\ Lab := complete\ AF\ Lab\ \wedge$$
$$(\forall L.complete\ AF\ L \wedge in(Lab) \subseteq in(L) \longrightarrow in(L) \approx in(Lab)).$$

The following well-known result can easily be verified:

$$grounded\ AF\ Lab = least\ (complete\ AF)\ Lab\ in$$

We now turn to complete labellings in which *undec-sets* must satisfy some minimality requirements: stable and semi-stable labellings.

Definition 3.9 [Stable labelling; cf. [1, Def. 24]] A labelling $\mathcal{L}ab$ is termed *stable* if it is a complete labelling whose *undec-set* is empty, i.e., no argument is labelled **Undec**.

The definition above is encoded in Isabelle/HOL as follows:

$$stable\ AF\ Lab := complete\ AF\ Lab \wedge (\forall x.Lab(x) \neq \mathtt{Undec})$$

Definition 3.10 [Semi-stable labelling; cf. [1, Def. 26]] A labelling $\mathcal{L}ab$ is termed *semi-stable* if it is a complete labelling whose *undec-set* is minimal (wrt. set inclusion) among all the complete labellings.

This definition is encoded in Isabelle/HOL as:

$$semistable\ AF\ Lab := minimal\ (complete\ AF)\ Lab\ undec$$

A complete overview of the encoding of labelling-based semantics in Isabelle is displayed in Fig. A.1 in Appendix B; when disregarding comments and further technical set-up, it only consists of less than 40 lines of code.

We have verified several (meta-)theoretical results for different sorts of complete (preferred, grounded, stable, semi-stable, ideal) labellings in theory `tests.thy`, which we cannot discuss here due to space limitations. In a similar vein, we have encoded and assessed the notions of ideal and stage labellings, and some notions for *skeptical* and *credulous* argument justification (cf. [1, §4]).

4 Flexible Generation of Labellings

The encoding for argumentation semantics presented above captures the structure and the logical behavior of argumentation frameworks within extensional type theory. Building on top of that, we can make use of automation tools from within Isabelle/HOL for generating labellings for concrete argumentation

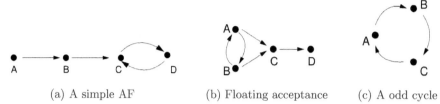

(a) A simple AF   (b) Floating acceptance   (c) A odd cycle

Fig. 1. Examples of argumentation frameworks. Graphics by Baroni, Caminada and Giacomin [1, Fig. 4-6].

frameworks. Figure 1 displays a few examples of argumentation frameworks taken from [1] that serve as use cases to illustrate the generation of labellings.

As an example, consider the argumentation framework from Fig. 1a: As a first step we define a new datatype, say $\mathtt{Arg}$, for its arguments – consisting only of the distinct terms $\mathtt{A}$, $\mathtt{B}$, $\mathtt{C}$ and $\mathtt{D}$. Next, we encode the attack relation $\mathtt{att}$ as binary predicate such that $\mathtt{att}\ X\ Y$ if and only if X attacks Y as displayed in Fig. 1a. The original source of this setup in Isabelle/HOL are displayed in Figure B.1 in Appendix B. We can now use the higher-order model finder Nitpick to ask for a, say, stable labelling. Indeed, Nitpick produces the following labelling (see the original output in Fig. B.2 in Appendix B):

$$Lab = x \mapsto \begin{cases} \mathtt{In} & \text{if } x = A \\ \mathtt{Out} & \text{if } x = B \\ \mathtt{Out} & \text{if } x = C \\ \mathtt{In} & \text{if } x = D \end{cases}$$

which represents the labelling Lab such that $in(Lab) = \{A, D\}$, $out(Lab) = \{B, C\}$ and $undec(Lab) = \emptyset$. Even more, we can employ Nitpick to generate all such labellings by just inspecting the value given to the free variable $LabSet$ in the formula below (cf. also Fig. B.3).

$$\forall Lab.\ LabSet\ Lab \longleftrightarrow stable\ att\ Lab$$

The reported labellings are in fact exactly those described in [1]. The same holds for the remaining argumentation semantics and examples from Fig. 1.

In addition to the above – quite standard – applications, we can now make use of the expressive surrounding logical framework to ask for *specific* labellings, e.g., satisfying an arbitrary (higher-order) predicate P. Consider the following example from Isabelle/HOL, relating to the example from Fig. 1b:

```
(* Admissible labelling where A is In *)
lemma ‹admissible att Lab ∧ Lab(A) = In› nitpick[satisfy] oops

(* Admissible labelling where Lab is surjective *)
lemma ‹admissible att Lab ∧ (surjective Lab)› nitpick[satisfy] oops

(* Admissible labelling where there are more than two arguments labelled In *)
lemma ‹admissible att Lab ∧ (card({x. in(Lab) x}) > 2)› nitpick[satisfy] oops
```

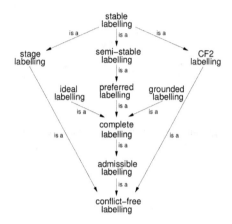

Fig. 2. An overview of the inclusion relations among the well-known labelling semantics. Graphics by Baroni, Caminada and Giacomin [1, Fig. 12].

In the three lemmas, we ask Nitpick to generate admissible labellings where, additionally, (1) argument A is labelled In, (2) Lab is a surjective function, and (3) there are more than two arguments labelled In, respectively. In the first two cases, suitable labellings are provided, in the third case no such labelling can be found (visualized by the red background color indicating an error). Indeed, no such labelling exists.

Similarly, we can prove in Isabelle/HOL that for Fig. 1c no admissible labelling other than the trivial one exists. This is expressed by the formula

$$admissible\ att\ Lab \longrightarrow \forall x.\ Lab(x) = \texttt{Undec}$$

which is proven automatically by Sledgehammer within a few seconds.

5   Application Perspectives

Verification of (Meta-)Logical Properties. In §3 we have mentioned, in passing, some of the results obtained through the use of Isabelle's automated tools (proven theorems and generated counter-models). In fact, the expressivity of HOL allowed us to formalize definitions and theorems involving general high-level mathematical notions such us minimality/maximality, least/greatest, fixed-points, etc., as well as domain-specific notions (e.g. *legally in/out*). These notions can, in turn, be used as building blocks to generate and assess further meta-logical properties (e.g., that for complete labellings in-sets completely determine the labelling). As an illustration, many of the inclusion relationships among the different labelling-based semantics, cf. Fig. 2, have been automatically verified in Isabelle/HOL using Sledgehammer [14, theory `tests.thy`].

Empirical Theory Exploration. In line with the above, a further contribution of the presented work consists in providing computer-assisted support for the (meta-)theoretical analysis of argumentation semantics. Next to verifying well-known results from the literature, (higher-order) theorem proving technology provides expressive means to formalize (meta-)theoretical conjectures and put them to the test by employing the reasoning tools of, e.g., Isabelle/HOL. This way we can explore the conceptual space of a theory in an empirical fashion, while interactively receiving feedback from the software about the adequacy of the formal model (e.g., by generating counterexamples).

Rich Instantiations of Argumentation Frameworks. There exist extended notions of abstract argumentation frameworks in the sense that the arguments are regarded as interpreted structures (as opposed to abstract objects), cf. [3] and the references therein. Using the here presented approach, we can easily interpret the abstract type 'a as complex structures within ExTT and, in turn, instantiate the argumentation framework with these objects. This includes, but is not limited to, taking arguments to be (sets or theories of) modal logic formulas, paraconsistent formulas or deontic logic formulas. For each of these examples there already exist embeddings into higher-order logic and, hence, a combination of these concepts is mostly engineering work.

6   Conclusion

In this paper, we modeled argumentation frameworks as typed binary relations in extensional type theory. Using this modeling, we provided an encoding of the well-known argumentation semantics, both in the extension-based and the labelling-based variant. The presented encoding was implemented using the Isabelle/HOL proof assistant, in which also selected experiments have been illustrated, including the generation of labellings and the verification of meta-logical properties.

The proposed approach is by no means meant (nor able) to challenge well-established and highly efficient procedures for calculating labellings (or extensions) of argumentation frameworks. It rather contributes a novel method of exploring the generation and analysis of argumentation semantics in the context of an expressive logical setting which cannot be addressed by existing approaches.

We argue that, mostly, the encoding re-formulates the basic definitions of argumentation semantics (without any major conceptual modifications) into the formal system of ExTT; and hence a formal proof would be mostly technical. Additionally, the logical and meta-logical experiments yielded the expected results in all cases. Still, a formal proof of faithfulness of the encoding would give a proper grounding of this work. Such a proof is further work.

The proposed approaches are, from a methodological point of view, in line with recent work in the computer-assisted assessment of normative theories using the LogiKEy [5] framework. Combining the here presented approaches and the formalized LogiKEy theories (also conducted in Isabelle/HOL), argumentation networks could be flexibly employed in the definition of ethico-legal governors. Another interesting path would be to incorporate explanation semantics [16] into the presented framework.

We encourage the interested reader to carry out further experiments, and also to further expand and improve on this work.

References

[1] Baroni, P., M. Caminada and M. Giacomin, *An introduction to argumentation semantics*, Knowl. Eng. Rev. **26** (2011), pp. 365–410.

[2] Baroni, P., D. M. Gabbay, M. Giacomin and L. van der Torre, "Handbook of formal argumentation," College Publications, 2018.

[3] Baroni, P., F. Toni and B. Verheij, *On the acceptability of arguments and its fundamental role in nonmonotonic reasoning, logic programming and n-person games: 25 years later*, Argument Comput. **11** (2020), pp. 1–14.

[4] Benzmüller, C., *Universal (meta-)logical reasoning: Recent successes*, Science of Computer Programming **172** (2019), pp. 48–62.

[5] Benzmüller, C., X. Parent and L. W. N. van der Torre, *Designing normative theories for ethical and legal reasoning: LogiKEy framework, methodology, and tool support*, Artif. Intell. **287** (2020), p. 103348.

[6] Benzmüller, C. and P. Andrews, *Church's Type Theory*, in: E. N. Zalta, editor, *The Stanford Encyclopedia of Philosophy*, Metaphysics Research Lab, Stanford University, 2019, summer 2019 edition .

[7] Blanchette, J. C., S. Böhme and L. C. Paulson, *Extending sledgehammer with SMT solvers*, J. Autom. Reason. **51** (2013), pp. 109–128.

[8] Blanchette, J. C. and T. Nipkow, *Nitpick: A counterexample generator for higher-order logic based on a relational model finder*, in: M. Kaufmann and L. C. Paulson, editors, *ITP 2010*, LNCS **6172** (2010), pp. 131–146.

[9] Caminada, M. W. A., W. A. Carnielli and P. E. Dunne, *Semi-stable semantics*, J. Log. Comput. **22** (2012), pp. 1207–1254.

[10] Caminada, M. W. A. and D. M. Gabbay, *A logical account of formal argumentation*, Stud Logica **93** (2009), pp. 109–145.

[11] Church, A., *A formulation of the simple theory of types*, J. Symb. Log. **5** (1940), pp. 56–68.

[12] Dung, P. M., *On the acceptability of arguments and its fundamental role in nonmonotonic reasoning, logic programming and n-person games*, Artif. Intell. **77** (1995), pp. 321–358.

[13] Dung, P. M., P. Mancarella and F. Toni, *Computing ideal sceptical argumentation*, Artif. Intell. **171** (2007), pp. 642–674.

[14] Fuenmayor, D. and A. Steen, *Isabelle/HOL sources associated with this paper*, Online available at Github: https://github.com/aureleeNet/formalizations (2021).

[15] Henkin, L., *Completeness in the theory of types*, J. Symb. Log. **15** (1950), pp. 81–91.

[16] Liao, B. and L. van der Torre, *Explanation semantics for abstract argumentation*, in: H. Prakken, S. Bistarelli, F. Santini and C. Taticchi, editors, *Computational Models of Argument - Proceedings of COMMA 2020, Perugia, Italy, September 4-11, 2020*, Frontiers in Artificial Intelligence and Applications **326** (2020), pp. 271–282.

[17] Nipkow, T., L. C. Paulson and M. Wenzel, "Isabelle/HOL - A Proof Assistant for Higher-Order Logic," Lecture Notes in Computer Science **2283**, Springer, 2002.

[18] Schulz, S., *E – a brainiac theorem prover*, AI Commun. **15** (2002), pp. 111–126.

[19] Steen, A. and C. Benzmüller, *Extensional higher-order paramodulation in Leo-III*, Journal of Automated Reasoning (2021).

[20] Wenzel, M., *Isabelle/Isar—a generic framework for human-readable proof documents*, From Insight to Proof-Festschrift in Honour of Andrzej Trybulec **10** (2007), pp. 277–298.

Appendix

A   Isabelle Formalization of Argumentation Frameworks

Fig. A.1. Overview of the Isabelle/HOL implementation of labelling-based argumentation semantics. The whole implementation consists of less than 40 lines of code.

B   Isabelle Setup for Extensions/Labellings Generation

```
(* Example set-up from [BCG2011], Figure 4 *)

(* A datatype introduces a new type for terms, here a finite number of
   distinct objects. Axioms stating the exhaustiveness and distinctness
   are generated automatically.  *)
datatype Arg = A | B | C | D

(* 'att' is defined as a term of type 'Arg Rel' (relation on objects of type 'Arg'),
   in fact as a predicate. The last line is short-hand syntax for defining
   the result of att as False for every pair of arguments not covered by the
   first four lines. *)
fun att :: ‹Arg Rel› where
   "att A B = True"
 | "att B C = True"
 | "att C D = True"
 | "att D C = True"
 | "att _ _ = False"
```

Fig. B.1. Encoding of the argumentation framework from Fig. 1a.

```
(* Ask Nitpick for a term Lab (a labelling) such that the predicate
   stable att Lab evaluates to true, i.e., such that Lab is a stable labelling for
   the argumentation framework represented by att.*)
lemma ‹stable att Lab› nitpick[satisfy] oops
◂

Nitpicking formula...
Nitpick found a model:

  Free variable:
    Lab = (λx. _)(A := In, B := Out, C := Out, D := In)
  Skolem constants:
    λx. ??.legallyIn.y = (λx. _)(A := A, B := A, C := A, D := A)
    λx. ??.legallyOut.y = (λx. _)(A := A, B := A, C := D, D := A)
  Types:
    Label = {In, Out, Undec}
```

Fig. B.2. Nitpick output for a stable labelling of argumentation framework in Fig. 1a.

```
(* Ask Nitpick for all labellings such that the predicate
   stable att Lab evaluates to true. This is done via an auxiliary predicate findFor2
   that defines a set of all labellings Labs.*)
lemma ‹findFor2 att stable Labs› nitpick[satisfy,box=false, eval = "card Labs"] oops
(* checked: these are exactly the two labellings given in [BCG2011] *)
◂
                        ☑ Proof state  ☑ Auto update   Update   Search:
Nitpick found a model:

  Free variable:
    Labs =
      {(λx. _)(A := In, B := Out, C := In, D := Out), (λx. _)
       (A := In, B := Out, C := Out, D := In)}
  Evaluated term:
    card Labs = 2
```

Fig. B.3. Nitpick output enumerating all stable labellings of the argumentation framework from Fig. 1a. For this we employed (an optimized variant of) the utility function: $findFor\ AF\ Prop\ S\ :=\ \forall Lab.\ S(Lab) \longleftrightarrow Prop(AF)\ Lab.$

Experiments in Causality and STIT

David Streit [1]

University of Luxembourg

Abstract

We present a framework for logical experimentation in Isabelle/HOL. We embed a STIT logic proposed by Marek Sergot in Isabelle/HOL and verify some of its properties computationally. We then propose a way to use computational tools to automatically check how different ways to define notions of causal responsibility behave in various cases. This paper is thus an exercise in experimental and computer-assisted logic.

Keywords: STIT, LogiKEy, actual causation, HOL, Isabelle, shallow embedding, experimentation.

1 Introduction

Logicians, like most people, are lazy. In this paper we present a methodology to help logicians steer their laziness towards a productive avenue.

We show how experimental, computer-assisted methods can be used to develop and understand causation in the context of STIT logics. To do so, we embed a STIT logic in Higher-Order-Logic (HOL) [2] using the proof assistant Isabelle/HOL [13]. We then model various example cases where causation plays a role in ethical deliberation in this STIT logic and show that computer-assisted experimentation can help tease out which parts of causation can be modeled using regular STIT logic and which parts cannot. Much of this is based on a paper by Marek Sergot [15] who proposed a way to introduce causality in STIT without modifying the underlying logic. [3]

Attribution of causation plays a role in both ethical and legal deliberation. However, its connection to STIT logic has only begun to be discussed. To the authors' knowledge, only two peer-reviewed papers [3,15] on the topic have been published, both of them less than one year old. Furthermore, for progress towards proper machine-assisted legal reasoning or even "machine ethics" notions of actions (like a STIT logic provides) and attribution of causality are indispensable. With this paper we try to make headway towards these goals.

---

[1] david.streit@uni.lu

[2] HOL is a typed higher order logic developed by Church in 1940. For further details see the original paper [10].

[3] Sergot elaborates these views in a yet to be published longer version [14].

The paper is structured in three parts. First, we will introduce the LogiKEy methodology [5], which is the tradition in which we see ourselves and show how one can embed logics in HOL in order to use already existing computational methods and how they can be incorporated in theory development.

Next, we introduce the logic we will embed. The logic of interest is a weakened standard STIT logic and its connection to causality was first studied by Marek Sergot [15]. We verified parts of his paper computationally.

In the final part, we present our embedding in Isabelle/HOL and show how automation can be used to quickly and efficiently try out various ways to introduce causal predicates. We show how to model example cases and then automatically get results how potential definitions of "is the cause of" or "could have prevented" behave in these cases. Since there is little consensus yet how one can (or cannot) define these notions in STIT at all, this serves as a tool to get an overview of the advantages and disadvantages for various proposals. For the reader to get a glimpse of why automation is fruitful here, consider the following back-of-the-envelope calculation. Suppose we have 5 potential definitions of "is the cause of" and we want to see how they behave in 5 different cases. Suppose further, that we already know what the intuitive outcome is in all these cases. With pen-and-paper, we would have to prove or disprove 25 statements. For some of those, we would first have to discover *whether* they hold or not.

With computational assistance, we can – in the best case – input the cases and definitions and let the computer calculate if the desired outcome holds for a given case and definition. In fact, this set-up is highly scalable. If we want to test yet another case or definition, we can simply add it and let the computer calculate.

2 The LogiKEy Methodology

Automated theorem proving for deontic logics has a problem: it does not scale with the invention of new logics. There is no agreed upon set of deontic logics. Instead, much of recent work in deontic logics includes proposing new logical systems in order to model specific cases. Numerous provers exist for propositional logic and first order predicate logic. For the vast majority of other logics, many of those deontic logics, no dedicated prover exists at all. Most logicians proposing a logical system have neither the time nor skills nor the desire to additionally write a prover for it.

This problem inspired the development of the LogiKEy framework. It uses HOL as a universal metalogic to embed other logics in [5]. HOL was chosen because it is both more expressive than first-order logic and dedicated high-performing provers exist. A logic embedded in HOL can thus be tackled by any prover that can solve HOL formulas (or relevant fragments thereof).

In the LogiKEy framework Henkin semantics for HOL are used, giving a complete semantics that still allows the faithful embedding of other logics [5]. The goal is to develop a system that can translate each formula ϕ in the target logic **L** to a formula $\lfloor \phi \rfloor$ in HOL such that $\models_{\mathbf{L}} \phi$ if and only if $\models_{HOL} \lfloor \phi \rfloor$. If this condition holds, we call the embedding *faithful*.

Paulson and Benzmüller [6] showed the faithfulness of embedding (multi-)modal normal logics in HOL in the manner used here. Restrictions of space prevent us from rehearsing their results here.

This is a general result for (quantified) normal modal logics in Paulson and Benzmüller [6]. The only difference to the logic introduced here is the additional fact that there is one accessibility relation $\sim$ that is a superset of all other accessibility relations (both those of individual agents and the resulting relations for groups). This fact [4] can be added at every relevant step in the proofs of Paulson and Benzmüller to show the faithfulness of the embedding here. In fact, the proofs can be simplified, since we are only interested in the non-quantified fragments here.

So every theorem proven in the embedding is in fact a theorem of Sergot's STIT logic. Faithful and scalable embeddings have also been proposed for Dyadic Deontic Logics [4] and others.

The LogiKEy methodology allows a user to combine and experiment with different logics at the same time. An embedding can be loaded like a library in a programming language to afford the user seamless access to different logics. At the inception of LogiKEy the authors focused on deontic logics, but the principle is available to many other unrelated fields. Additionally, so far, no logic dealing with actions or causation has been included in the LogiKEy framework, making the present project a potential addition.

Embedding a logic in Isabelle/HOL in this manner allows one not only to make use of the existing provers for HOL, but also enables the integrated tool support of Isabelle/HOL. The two most important tools for the present purpose are the countermodel finder *Nitpick* and the proof routine *Sledgehammer*. Sledgehammer tries to automatically find prove tactics for a given statement. It makes use of machine learning approaches to find appropriate lemmas and statements that have already been verified in its attempt to generate proofs.

Contrary to most current uses of the LogiKEy framework, where the main purpose is to use existing logics as a framework to enable the user to formalize and reason about extra- or metalogical issues, we primarily use the embedding to show how it can be used for *intra-logical* theory development. Another adjacent paper uses the LogiKEy framework in helping developing a logic for value based legal reasoning [9].

3 Sergot's STIT Logic

The logic we are embedding is a normal multimodal logic. Its main modal operator $[G]$ models the standard STIT predicate.[5] It can be read as "(the group of agents) G see to it that". We assume that the set of agents (*"Ag"*) is

---

[4] and the "translation" of this fact to Henkin models, where we need to guarantee that $D_{\mu \to \mu \to o}$ still has this property and the resulting model is still a Henkin model. For more details, see the proofs in the original paper.

[5] It is what other authors, most prominently Horty [11] often refer to as the Chellas-stit or cstit.

finite and we have a STIT operator for each subset $G \subseteq Ag$. Furthermore, we have an operator $\Box$ for "historical necessity" or "settledness" as in other STIT logics. $\Box$ corresponds to the STIT operator for the empty set of agents $[\emptyset]$.[6]

Thus the BNF for the logic can be given as:

$$\mathbf{p}|\bot|\neg\mathbf{p}|\mathbf{p} \vee \mathbf{p}|\mathbf{p} \wedge \mathbf{p} \to \mathbf{p}|\Box\mathbf{p}|[\mathbf{G}]\mathbf{p} \text{ for every } G \subseteq Ag$$

The axiom schemes of this logic are the following:
$\Box$ and G are S5 operators for every $G \subseteq Ag$ and

$$\text{If } G \subseteq H, \text{ then } [G]\phi \to [H]\phi$$

The latter axiom will be called "superadditivity". Sound and complete semantics for this logic can be given via the following [15]:

Let $M = (W, \sim, \sim_{x \in Ag}, V)$ where W is a non-empty set of worlds and V is a valuation function. We further demand that $\sim$ and $\sim_x$ are equivalence relations for every $x \in Ag$. Lastly, $\sim_x \subseteq \sim$ for every $x \in Ag$. As is tradition in Kripke semantics, we call these accessibility relations.

We then lift the relations to subsets of Ag. Let $G \subseteq Ag$, then $\sim_G = \sim \cap \bigcap_{x \in G} \sim_x$. With this, we define satisfiability as usual, with satisfiability for the modal operators being defined as:
$\Box\phi$ *holds in a world* τ *iff for every word* τ' *such that* $\tau \sim \tau'$: ϕ *holds in* τ'
$[G]\phi$ *holds in a world* τ *iff for every world* τ' *such that* $\tau \sim_G \tau'$: ϕ *holds in* τ'

The logic introduces here corresponds to standard STIT axioms without "independence of agents".

Causation in STIT With this logic in place we can model outcomes of actions of groups of agents. But often this is not enough to model ethical or legal situations. STIT easily generalizes to groups of agents. Imagine a case where two agent's both give someone half a lethal dose of poison. It is intuitive to say that as a group they caused the outcome, even though none of the individual actions were sufficient. As another example, think of the case of the vase, which is also present in Sergot's paper [15].

There is a precious vase in the house of Alice and Bob. Both Alice and Bob can perform an action to put the vase outside. There is also the possibility (either a "real" possibility or a possibility relative to some salient context) that it rains. If it rains the vase gets wet and is ruined.

In simple multi-agent systems causal responsibility of a group G for ϕ is usually taken to include three elements. (1) ϕ must obtain. (2) ϕ could have been prevented by G and (3) G is minimal.[7] Now assume that Alice takes the vase

---

[6] Both because it allows a more efficient embedding and to stay consistent with Sergot's paper we still include $\Box$ as a primitive here.

[7] Not all of these are present in every case. Naumov et al. [12] for example does not include minimality. How (3) is to be understood is usually the most contentious claim. One possibility, see below, is to simply use $\Diamond[G]\neg\phi$. This idea is close to what Baltag et al. [3] but also Yazdanpanah et al. [16] propose. It can however lead to unintuitive results when it comes to chancy causation where the outcome is not guaranteed. In some cases (for example

outside and it does in fact rain and the vase is ruined. Clearly there is a sense in which Alice (but not Bob) is responsible for it. But this is not easily captured in STIT logic. First, note that it is not the case that $[\{Alice\}]VaseIsWet$, since there is a world where it did not rain and thus the vase is not ruined. It is also not the case that Alice could have guaranteed that the vase is not wet. So $\Diamond[Alice]\neg VaseIsWet$ is also not true, since Bob could have taken the vase outside.

This invites the question, if STIT logic is simply missing important concepts to model this kind of responsibility. Why is it interesting to focus on STIT here instead of developing an explicit logic of responsibility or blame (see [12,16])? For one thing, STIT is an action logic that can model some cases quite nicely. If it were the case that we can define causal predicates in this logic we would "get them for free". In his paper Marek Sergot argues that, indeed, large swaths of cases of responsibility attribution can in fact be modeled in STIT logic.

We will try to tackle this problem with computational assistance.

4 Embedding & Results

The embedding in Isabelle/HOL follows the strategy proposed earlier and is very closely modeled on work by Paulson and Benzmüller [6]. The source code of the embedding can be found online [1] [8].

First, we define a type for possible worlds and a type for agents which we require to be finite.

typedecl i — Type for worlds
typedecl μ — Type for agents

abbreviation $"Ag \equiv (UNIV::\mu\ set)"$ — every element of type μ is an agent
axiomatization where $finiteActors:$ $"finite\ Ag"$ — The set of agents is finite

Next we introduce a derived type that will represent formulas. These are of the type $i \to bool$. Intuitively these represent truth in a world.

type_synonym $\sigma = "(i \Rightarrow bool)"$ — type for formulas

We can then implement our accessibility relations as constants in HOL and define the requirements on them.

abbreviation $"reflexive\ R \equiv (\forall x.\ R\ x\ x)"$
abbreviation $"symmetric\ R \equiv (\forall x\ y.\ R\ x\ y \longrightarrow R\ y\ x)"$
abbreviation $"transitive\ R \equiv (\forall x\ y\ z.\ (R\ x\ y) \wedge (R\ y\ z) \longrightarrow (R\ x\ z))"$

---

the jumper and shooter case (see [14]) it gives arguably the superior outcome. Testing in Isabelle can confirm this "false negative" finding of Sergot's preferred definition.

[8] The source code is seperated in 5 Files. First the embedding in *Sergot.thy*, second some useful lemmas, often times verifying results from Sergot's paper in *SergotTheorems.thy*, we then have the various cases modeled as locales in *SergotCases.thy* and possible definitions of *Could* in *SergotCould.thy* and lastly the experiments as described below in *SergotTests.thy*.

abbreviation *"eqrelation R ≡ reflexive R ∧ symmetric R ∧ transitive R"*

consts *tilde* :: *"i⇒i⇒bool"* (**infixr** *"˜"* 70) — accessibility relation (global)
consts *tildeAG* :: *"i⇒μ⇒i⇒bool"* (*"_ ˜_ _"*) — acc. rel. (agent relative)

axiomatization where *xsub:* *"∀ (x::μ). ((τa::i)˜x(τb::i) ⟶*
((τa::i)˜(τb::i)))" — acc. rel. are subsets
axiomatization where *tildeeqU:* *"∀x y. x˜y"* — S5U acc. rel.

Note one slight departure from Sergot's original semantics. Instead of simply having ∼ be a equivalence relation, we demand that every world is accessible from every world. This corresponds to the semantics for S5U. These stronger semantics are however equivalent to Sergot's.[9] Using them makes the embedding simpler and thus automation faster.

Logical connectives are then defined using lambda notation. They are given as abbreviations, meaning that Isabelle automatically unfolds these definitions internally, but the user can input them as if they were undefined objects. For brevity only three of these are given here. To differentiate these defined operators from the metalogic operators of HOL, they will be typeset in bold font.

abbreviation *mor*   :: *"σ⇒σ⇒σ"* (**infixr** *"∨"* 50)
    where *"φ∨ψ ≡ λw. φ(w)∨ψ(w)"*
abbreviation *mstit* :: *"μ set ⇒ σ⇒σ"* (*"[_] _"* [52]53)
    where *"[G] φ ≡ λ(w::i).∀ (v::i). (w ˜ G v) ⟶ φ(v)"*
abbreviation *mstitDIA*   :: *"μ set ⇒ σ⇒σ"* (*"<_> _"* [52]53)
    where *"<G> φ ≡ λ(w::i).∃ (v::i). (w ˜ G v) ∧ φ(v)"*
axiomatization where *tildexeq:* *"∀ (x::μ). eqrelation (λa b.*
((a::i)˜x(b::i)))"
abbreviation *tildeG* :: *"i⇒ (μ set)⇒i⇒bool"* (*"_ ˜_ _"*)
    where *"(τa ˜G τb) ≡ (τa˜τb) ∧ (∀x. (x∈G⟶ ((τa::i)˜x(τb::i))))"*
— acc. rel. for groups is intersection of acc rel. for agents

Finally, we can define validity in a world and validity with respect to the class of models we are after.

abbreviation *valid* :: *"σ⇒bool"* (*"⌊_⌋"* [8]109)
    where *"⌊p⌋ ≡ ∀w. p w"*
abbreviation *follows_w* :: *"i ⇒ σ ⇒ bool"* (**infix** *"⊨"* 55)
    where *"(w ⊨ p) ≡ p w "*

With the semantics in place, we can prove object language formulas in Isabelle. Easier formulas can be proven automatically by internal or external provers.

For example, the prover can automatically verify that the axiom scheme

---

[9] The proofs for soundness and completeness carry over once it has been established that S5 semantics using (just) equivalence relations and S5U semantics with a universal accessibility relation are both sound and complete with respect to S5 axioms.

superadditivity holds.

lemma `Superadditivity:` `"G ⊆ H ⟶ ⌊([G]φ) → ([H]φ)⌋"` **by** `auto`

Modeling Cases Before we can model cases, we need to be able to refer to different actions. Actions in standard STIT do not have names. In contrast to a branching-time semantics for STIT, we model an action as a set of worlds instead of as a set of histories. [10] A world is reachable from another world via $\sim_x$ iff it is in the same action (of an agent x).

Finally, the set of actions available to an agent in a world τ is simply the set of actions from above. We can encode this in Isabelle like this:

abbreviation `alt` `::` `"μ ⇒ i ⇒ i set"` `("alt _( _ )")`
    where `"alt x( τ ) ≡ {τb. (˜ x) τ τb}"`
abbreviation `actiontypes` `::` `"μ ⇒ i set set"` `("A _")`
    where `"A x ≡ {v. ∃τ. v = alt x (τ)}"`

Recall the motivating example of the vase. How can we model it in Isabelle? We will make use of Isabelle's ability to define "locales". For the present purpose, it is sufficient to think of them as blocks where certain (axiomatic) conditions hold. For further details, see the official documentation [2]. We can later prove theorems in specific locales.

We will create one locale for each case we model. Each case has a set of (sometimes implicit) assumptions that we encode in the locale. Inside the locale, each assumption (and nothing more) holds. So if we can prove or disprove a statement in a locale, we can think of it as being true/false in this specific case.

Once we have modeled several cases as locales, we can try to prove or disprove the same statement for several cases by simply switching the locale.

As an illustration we will model the case of the vase in Isabelle by defining a locale "Vase". First, we introduce two constants for the agents "a" and "b" and an action for them taking the vase outside. Additionally, we introduce constants intended to mean that the vase is inside/wet or that it is raining. [11]

In the next part, we specify the locale to match the case. We do this by giving constraints on what worlds are possible. For example, we want there to be a world where a takes the vase outside but not b, but no world where the vase is outside but neither a nor b brought it there. We hope the reader can easily match the rest of the conditions herself. To simplify the case a little we also add a condition that "not bringing the vase" is also an action.

---

[10] Sergot's semantics are more restrictive than other STIT semantics, but they can be extended to include time-indexed formulas. For the purpose of this paper however, we prefer a simpler logic, both for conceptual and computational reasons. Of course, this precludes studying strategies and causal responsibility in sequences of actions in STIT.

[11] We could also define them inside the locale, but since we use the same constants in different locales, we define them outside of any locale.

consts `abringsvase::"i set"` — the action of a bringing the vase outside
consts `bbringsvase::"i set"` — the action of b bringing the vase outside
consts `inside::"σ"` — the vase is inside
consts `rain::"σ"` — it is raining
consts `wet::"σ"` — the vase is wet

locale `Vase =`
assumes `"Ag = {a, b}"` — We only have two agents
and `"a ≠ b"` — they are different
and `"wet = ¬inside ∧ rain"` — the vase is ruined if it's outside and it rains
and `at: "abringsvase ∈ (A a)"` — abringsvase is an action of agent a
and `bt: "bbringsvase ∈ (A b)"` — bbringsvase is an action of agent b
and `"(UNIV::i set) - abringsvase ∈ (A a)"` — so is not bringing the vase
and `"(UNIV::i set) - bbringsvase ∈ (A b)"`
and `"∃τ. (τ ∈ abringsvase ∧ τ ∉ bbringsvase)"` — It is possible for a to bring the vase but not b
and `"∃τ. (τ ∉ abringsvase ∧ τ ∈ bbringsvase)"` — and vice verse
and `"¬ (∃τ. (τ ∈ abringsvase ∧ τ ∈ bbringsvase))"` — Both cannot bring the vase outside (optional)
. . .
and `"∀τ. (τ ∉ abringsvase ⟶ τ ∉ bbringsvase ⟶ τ ⊨ inside)"` — By default the vase is inside
and `"∃τ. (τ ∈ abringsvase ∧ τ ⊨ rain)"` — It might rain
. . .
and `"∃τ. (τ ∉ abringsvase ∧ τ ∉ bbringsvase ∧ τ ⊨ (¬ rain))"`— And so forth...

If we then want to prove a statement inside this locale we can simply append "(in Vase)" to a lemma in Isabelle. We formalized various cases taken from Sergot's paper in this way. With these in place, the last element needed for our testing suite is a plethora of plausible definitions of "being the cause of". For illustrative purposes we will start with Sergot's preferred definition. For an argument why this is a good candidate [12] see the paper [15].

abbreviation `Could :: "μ set ⇒ σ ⇒ σ"` `("Could _ _")`
    where `"Could G φ ≡ <Ag - G> [Ag] φ"`
abbreviation `Couldmin :: "μ set ⇒ σ ⇒ σ"` `("Could^{min} _ _")`
    where `"(Could^{min} G φ) (τ::i) ≡ ((Could G φ) τ) ∧ ¬(∃ H. (H ⊂ G)`
`∧ ((Could H φ) τ))"`

G being the actual cause of an outcome ϕ is then the formula $\phi \land Could_G^{min}\neg\phi$.
    The desired outcome for the vase situation is: If a brings the vase outside and it rains (and thus the vase is wet), then a is the actual cause of the vase being wet. So what we need to test in Isabelle is, if this statement holds:

---

[12] the main idea is: treat the predicate as "making a difference while holding other agent's actions fixed".

lemma (**in** Vase) "($\tau \in$ abringsvase) $\longrightarrow$ ($\tau \models$ wet) $\longrightarrow$ ($\tau \models$ (Could^{min} {a} ($\neg$ wet)))"

Isabelle offers two powerful tools to decide questions like this: Nitpick and Sledgehammer. Nitpick tries to generate countermodels for a given statement, while Sledgehammer automatically invokes various provers to generate a proof. To find out if a given statement is a theorem the user can use both in parallel. To ensure that the axioms of the logic are included we run Nitpick with the setting "user_axioms, timeout=300".

lemma (**in** Vase) "$\neg$ (($\tau \in$ abringsvase) $\longrightarrow$ ($\tau \models$ wet) $\longrightarrow$ ($\tau \models$ (Could^{min} {a} ($\neg$ wet))))" **nitpick sledgehammer**

If this does not lead to a result, we could also do the same for the negation of this statement

lemma (**in** Vase) "$\neg(\forall \tau.$ ($\tau \in$ abringsvase) $\longrightarrow$ ($\tau \models$ wet) $\longrightarrow$ ($\tau \models$ (Could^{min} {a} ($\neg$ wet)))" **nitpick sledgehammer**

In this case however, Sledgehammer will find a proof for the former statement and we have verified computationally that in this case, Sergot's definition does indeed give the desired result. In this example, Nitpick fails to find a countermodel even after a long search. In many other instances however, it is the (counter-)model finder Nitpick that provides the answer.

What other possible definitions of "could" are possible? Sergot offers one strong contender but also mentions others in the paper.

The first is to focus on "If an agent/a group of agents had done something different". Sergot formalizes this by using the complement of the accessibility relation. In Isabelle we can write it as:

abbreviation tildeGconv :: "i$\Rightarrow$ (μ set)$\Rightarrow$i$\Rightarrow$bool" ("_ -~-_ _")
  where "(τa -~- (G::μ set) τb) $\equiv$ (τa ~ τb) $\wedge$ ($\forall$x. ((x $\in$ G) $\longrightarrow$ $\neg$(τa::i)~x(τb::i)))"

With these, we can define two new operators. A STIT operator based on this converse accessibility and its "diamond" dual.

abbreviation mstitconv :: "μ set $\Rightarrow$ $\sigma \Rightarrow \sigma$" ("[[_]] _"[52]53)
  where "[[(G::μ set)]] $\varphi \equiv \lambda$(w::i).$\forall$ (v::i). (w -~- G v) $\longrightarrow$ φ(v)"
abbreviation mstitconvDia :: "μ set $\Rightarrow$ $\sigma \Rightarrow \sigma$" ("⟨ _ ⟩ _"[52]53)
  where "⟨(G::μ set)⟩ $\varphi \equiv \lambda$(w::i).$\exists$ (v::i). (w -~- G v) $\longrightarrow$ φ(v)"

Which in turn will simply function as alternative definitions for "could". Both times, we demand that the set in question be minimal.

abbreviation Could2 :: "μ set $\Rightarrow$ σ $\Rightarrow$ σ" ("Could2 _ _")

where *"Could2 G φ $\equiv$ $[\![(G::\mu$ set$)]\!]$ (φ)"*
abbreviation *Could3* :: *"μ set $\Rightarrow$ σ $\Rightarrow$ σ"* *("Could3 _ _")*
where *"Could3 G φ $\equiv$ $\langle\!\langle(G::\mu$ set$)\rangle\!\rangle$ φ"*

The last option we mention, is simply the ability to prevent an outcome from happening. This also has some intuitive appeal. In fact, this idea is behind proposals like Baltag et al..[13]

abbreviation *Could4* :: *"μ set $\Rightarrow$ σ $\Rightarrow$ σ"* *("Could4 _ _")*
where *"Could4 G φ $\equiv$ $\Diamond$ $[G]\varphi$"*

For all these we also define their minimal counterparts. Sergot mentions other possibilities he ultimately rejects. We could also come up with new definitions or use combinations of the ones already introduced. But since the aim of this paper is mostly one of methodology, we restrict ourselves to these four.

For the single case we thus have eight statements of interest:

lemma (**in** *Vase*) *"(τ $\in$ abringsvase) $\longrightarrow$ (τ $\models$ wet) $\longrightarrow$ (τ $\models$ (Could^{min} {a} ($\neg$ wet)))"* **nitpick sledgehammer**
lemma (**in** *Vase*) *"$\neg$($\forall\tau$.(τ abringsvase) $\longrightarrow$ (τ $\models$ wet) $\longrightarrow$ (τ $\models$ (Could^{min} {a} ($\neg$ wet))))"* **nitpick sledgehammer**
. . .
lemma (**in** *Vase*) *"(τ $\in$ abringsvase) $\longrightarrow$ (τ $\models$ wet) $\longrightarrow$ (τ $\models$ (Could4^{min} {a} ($\neg$ wet)))"* **nitpick sledgehammer**
lemma (**in** *Vase*) *"$\neg$($\forall\tau$.(τ $\in$ abringsvase) $\longrightarrow$ (τ $\models$ wet) $\longrightarrow$ (τ $\models$ (Could4^{min} {a} ($\neg$ wet))))"* **nitpick sledgehammer**

If we wanted to automatically test a candidate definition for "could" we could simply add two lines of code. If we wanted to test these definitions with a different case of interest we can – after defining the case as a locale as we did with the vase case – simply copy & paste the lines above and replace "in Vase" with the respective locale of the case one has added.

From a user's perspective, this scales more easily than doing this with pen-and-paper. Of course, coming up with relevant cases or definitions of "could" is still a task that requires at least an initial understanding of the logic, but the user is in a situation where she can simply come up with an idea and leave it to the computer to find out if it behaves in the way that she suspects.

Results Naturally, this is only of interest, if the computer is actually able to generate counterexamples/proofs in a large number of cases. Otherwise, it might be faster to simply try to prove the statements by hand, instead of going the extra mile to input the cases and definitions in the first place. So how are the results empirically?

---

[13] In the vase example and other cases of chancy causation, this gives not the desired result. In many cases – some that are not covered by Sergot's proposal – it does in fact provide the desired outcome.

We have modeled four cases taken from Sergot. For each case we formulated a statement that corresponds to the "intuitive result". We then tried to prove this statement via Sledgehammer and tried to find a counterexample via Nitpick. We did the same for the negation of the statement. In total, we ran Nitpick and Sledgehammer 32 times. We set the timeout for both to 5 minutes. In the majority of cases, if a prove/countermodel was found, it was found considerably faster. All experiments were run on a commercial laptop from 2016.

We consider the test a success if either a counterexample for a statement or its negation was found or a statement or its negation could be proven. [14] A total of 11 of the 16 statements, were successful. Additionally, one further statement could be proven, with minimal human input. 4 of the 16 statements could not automatically be verified/refuted. These were clustered around two cases, so it might be that the structure of the cases was more challenging to provers.

The proof assistant confirmed Sergot's results and no major mistakes have been unearthed by the proof assistant.

How well automation performs is largely a function of the complexity of a statement that needs to be checked and how complex the embedding is. In this case, the embedding was of moderate complexity. The only major hurdle were statements and definition that used sets. Previous embeddings (e. g. [6]) made use of characteristic functions instead of sets to improve efficiency. We have found that removing sets in favor of their characteristic functions did *not* made proofs easier. It also made it less intuitive to introduce new definitions and model cases. [15]

Still, in roughly three out of four cases, the proof assistant was able to quickly provide a result. We suggest that automated proof assistants can be a fruitful tool in theory development.

5 Related and Future Work

Some of the proposals for different "coulds" were either underinclusive or over-inclusive. That is, in some cases they either attributed too little or too much causal responsibility. Usually a few cases are enough to exclude a proposed definition from further consideration. With the framework in place however, we

---

[14] For some of the proofs generated, the proof object could not easily be reconstructed or proof checking took a long time. Usually, in computer assisted proving this is a problem and requires human input. In our case however, we are simply interested in whether a given statement is true and do not need to be able to have a proof object reconstructed/checked in Isabelle.

[15] A reviewer suggested treating agents as datatypes and use injective mappings to type μ to avoid sets. We suspect that there are indeed performance improvements to be gained. Using sets however has the significant advantage that they are transparent to any user. We face a trade off between computationally faster encodings and *easy* encodings. Part of what we want to show is that a cursory familiarity with a theorem prover is enough to have a significant improvement over pen-and-paper. Therefore, we decided to include axiomatic definitions using sets.

are able to perform new experiments. Are combinations (via a disjunction for example) of underinclusive "coulds" still underinclusive? If so, is there a single case that can rule them out all in one fell swoop? Relatedly, are combinations of underinclusive "coulds" overinclusive in some cases? That is, is at least one of them not only underinclusive, but also overinclusive? If so, we can generate counterexamples more reliably.

We can also do the converse. Are combinations (for example via a conjunction) of overinclusive "coulds" still overinclusive? We hope that these experiments might be fruitful in finding differences in meaning for different concepts of causal attribution. Differences of this sort in causal STIT have been proposed both by Sergot [15] and Batlag et al. [3], however we think that an experimental framework might generate new insights, due to the sheer number of combinations of cases and definitions it can oversee.

While, to the knowledge of the author, this is the first attempt to generate a testing environment for possible definitions for a concept where so far, no consensus has been reached, automated theorem proving have been used to find better premises for philosophical arguments [8], verify and improve textbooks [7] and reason in a variety of logics before.

With regard to STIT logics and causation Baltag et al. proposed a more complicated semantics [3]. In their paper, they use named actions and an opposing relation between them. This is in turn used to define an "expected result", a result that necessarily holds if the action is done unopposed. They combine this notion with counterfactual tests to get at a notion of causal responsibility. There is a lot of theoretical overlap between the proposal by Sergot and Baltag et al. that we do not have space to address here. A potential fruitful avenue is to use the testing suit to see if there is a privileged set of cases that can only be modeled by Baltag et al.'s proposal or if the simpler approach by Sergot might be sufficient.

Standard STIT semantics with action types are computationally more demanding and so far only simpler semantics have been implemented in Isabelle. We doubt that we could achieve a similar test environment for their semantics. However, we plan to model the cases that serve as motivation in their paper in an attempt to make it clearer what the advantages of a more complicated semantics are. We hope to be able to develop a class of cases where differences become visible and easily understandable. This work, will not be entirely automated, nor computerized of course. However, we hope to automate away the boring parts and allow the logician to focus on the more conceptual tasks in theory development.

References

[1] *https://github.com/lngaisubmission/submission.*

[2] Ballarin, C., *Locales and locale expressions in isabelle/isar*, in: *International Workshop on Types for Proofs and Programs*, Springer, 2003, pp. 34–50.

[3] Baltag, A., I. Canavotto and S. Smets, *Causal agency and responsibility: a refinement of stit logic*, in: *Logic in High Definition*, Springer, 2021 pp. 149–176.

[4] Benzmüller, C., A. Farjami and X. Parent, *Dyadic deontic logic in HOL: Faithful embedding and meta-theoretical experiments*, in: M. Armgardt, H. C. Nordtveit Kvernenes and S. Rahman, editors, *New Developments in Legal Reasoning and Logic: From Ancient Law to Modern Legal Systems*, Logic, Argumentation & Reasoning **23**, Springer Nature Switzerland AG, 2021 .

[5] Benzmüller, C., X. Parent and L. van der Torre, *Designing normative theories for ethical and legal reasoning: Logikey framework, methodology, and tool support*, Artificial Intelligence **287** (2020), p. 103348.

[6] Benzmüller, C. and L. C. Paulson, *Quantified multimodal logics in simple type theory*, Logica Universalis **7** (2013), pp. 7–20.

[7] Benzmüller, C. and D. S. Scott, *Automating free logic in HOL, with an experimental application in category theory*, Journal of Automated Reasoning **64** (2020), pp. 53–72.

[8] Benzmüller, C., *A (Simplified) Supreme Being Necessarily Exists, says the Computer: Computationally Explored Variants of Gödel's Ontological Argument*, in: *Proceedings of the 17th International Conference on Principles of Knowledge Representation and Reasoning*, 2020, pp. 779–789.

[9] Benzmüller, C. and D. Fuenmayor, *Value-oriented Legal Argumentation in Isabelle/HOL* (2021).

[10] Church, A., *A formulation of the simple theory of types*, The Journal of Symbolic Logic **5** (1940), pp. 56–68.

[11] Horty, J. F., "Agency and deontic logic," Oxford University Press, 2001.

[12] Naumov, P. and J. Tao, *Blameworthiness in strategic games*, Proceedings of the AAAI Conference on Artificial Intelligence **33** (2019), pp. 3011–3018.

[13] Nipkow, T., L. C. Paulson and M. Wenzel, "Isabelle/HOL: a proof assistant for higher-order logic," Springer Science and Business Media, 2002.

[14] Sergot, M., *Actual cause and chancy causation in 'stit'*, unpublished.

[15] Sergot, M., *Actual cause and chancy causation in 'stit': A preliminary account*, in: P. McNamara, A. J. I. Jones and M. A. Brown, editors, *Agency, Norms, Inquiry, and Artifacts: Essays in Honor of Risto Hilpinen*, Synthese Library, Springer, 2021 .

[16] Yazdanpanah, V., M. Dastani, W. Jamroga, N. Alechina and B. Logan, *Strategic responsibility under imperfect information*, AAMAS '19 (2019), p. 592–600.

A Brief Introduction to the Shkop Approach to Conflict Resolution in Formal Argumentation

Dov Gabbay [1]

*University of Luxembourg, Luxembourg and
King's College London, United Kingdom and
Bar Ilan University, Israel*

Timotheus Kampik [2]

Umeå University, Sweden

Abstract

In this paper, we formalise the *Shkop* approach to conflict resolution in formal argumentation, in which we start with an empty abstract argumentation framework *AF* and an initially empty set of inferred arguments. Then, we expand *AF* one argument at a time, and evaluate after each expansion if *i)* arguments that have previously been inferred can be kept (or have to be discarded due to sufficient doubt) and *ii)* if the newly added argument can be added to the set of inferred arguments. Based on this idea, we introduce a novel approach for designing abstract argumentation semantics. As a particular semantics, we define *grounded Shkop* semantics – a naive set-based argumentation semantics that does not inhibit a well-known problem of CF2 semantics.

Keywords: abstract argumentation, non-monotonic reasoning, argumentation semantics

1 Introduction

Over the past decades, formal argumentation has emerged as a promising collection of methods for non-monotonic reasoning. *Abstract* argumentation, a foundational variant of formal argumentation, encompasses approaches for drawing inferences from abstract argumentation frameworks, which are tuples of a set of abstract arguments and a binary relation (*attacks*) on these arguments [8]. Inferences are drawn using so-called *argumentation semantics*, functions that take an argumentation framework and return *extensions*, *i.e.*, subsets of the argumentation framework's arguments, as potential conclusions. The way an argumentation semantics should determine an argumentation framework's extensions depends on the application scenario and the intended meaning of arguments and attacks; hence, an important research direction in

---

[1] dov.gabbay@kcl.ac.uk

[2] tkampik@cs.umu.se. TK was supported by the Wallenberg AI, Autonomous Systems and Software Program (WASP) funded by the Knut and Alice Wallenberg Foundation.

abstract argumentation is the design and analysis of argumentation semantics. Many formal principles that analyse semantics behaviour have been defined [12], and several families of argumentation semantics exist [2]; for example, all extensions yielded by semantics in the *naive set-based* semantics family are $\subseteq$-maximal among conflict-free sets; hence, these semantics satisfy the *naivety* principle.

In this paper, we present the *Shkop* [3] approach to inferring conclusions from argumentation frameworks, starting with a sequential perspective on argumentation (Section 3). We extend the approach to create a new way to specify naive-based semantics that are based on an intuitive approach to conflict resolution, define *grounded Shkop* semantics, a particular *Shkop semantics* variant (Section 4), and provide a preliminary analysis that highlights some advantages that grounded Shkop has over some other naive set-based semantics, and in particular over CF2 semantics (Section 5). A comparison to the recently introduced SCF2 semantics [7] remains an open issue.

This paper is accompanied by an implementation that extends the DiArg dialogue reasoner [4] . The implementation is available at `https://git.io/JOiEF`.

2   Preliminaries

This section provides the theoretical preliminaries of our work, starting with the central notion of an (abstract) argumentation framework as introduced by Dung in his seminal paper [8]. An argumentation framework AF is a tuple (AR, AT), such that AR is a set of elements ("arguments") and $AT \subseteq AR \times AR$ ("attacks"). For $(a, b) \in AT$, we say that a attacks b. For $S \subseteq AR$ and $a \in AR$, if $\exists b \in S$ and $(b, a) \in AT$, we say that S attacks a; if $\exists c \in S$ and $(a, c) \in AT$, we say that a attacks S. For $S \subseteq AR$, $P \subseteq AR$ such that $\exists (a, b) \in AT, a \in S, b \in P$, we say that S attacks P; we denote all arguments attacked by S by S^{+} (the *range* of S) and all arguments that attack S by S^{-}. For any argument $c \in AR$ such that $(c, c) \in AT$, we say that c is a self-attacking argument. For $S \subseteq AR$, we say that S defends a iff $\forall d \in AR$, such that d attacks a, S attacks d.

Informally speaking, a restriction of an argumentation framework $AF = (AR, AT)$ to a set of arguments set S removes all arguments that are not in S from AR, and all attacks from or to arguments not in S from AT.

Definition 2.1 [Restriction [4]] Let $AF = (AR, AT)$ be an argumentation framework. Given a set $S \subseteq AR$, let $AF \downarrow_S$ be defined as $(S, AT \cap S \times S)$. We call $AF \downarrow_S$ the *restriction of AF to S*.

Often, the influence of self-attacking arguments on the behaviour of an argumentation semantics is not desirable. Hence, Cramer and Van der Torre provide an abstraction that removes all self-attacking arguments from a given argumentation framework [7].

Definition 2.2 [nsa(AF)] Let $AF = (AR, AT)$ be an argumentation framework. We define $nsa(AF) = AF \downarrow_{AR'}$, where $AR' = \{a | a \in AR$ and $(a, a) \notin AT\}$.

---

[3] The approach is based on semi-formally presented loop-busting methods in Talmudic logic and is named after Rabbi Shimon Shkop (1860–1930), a scholar who analysed logical principles in the Talmud [1].

[4] DiArg is based on the Tweety project's argumentation libraries [11].

Important concepts in abstract argumentation are the notions of conflict-free and admissible sets.

Definition 2.3 [Conflict-free and Admissible Sets [8]] Let $AF = (AR, AT)$ be an argumentation framework. A set $S \subseteq AR$:

- is *conflict-free* iff $\nexists a, b \in S$ such that a attacks b;
- is *admissible* iff S is conflict-free and $\forall a \in S$, it holds true that S defends a.

Let us now define the notion of a path between two arguments in an argumentation framework.

Definition 2.4 [Path between Arguments] Let $AF = (AR, AT)$ be an argumentation framework. A path from an argument $a_0 \in AR$ to another argument $a_n \in AR$ is a sequence of arguments $P_{a_0, a_n} = \langle a_0, ..., a_n \rangle$, such that for $0 \leq i < n$, a_i attacks a_{i+1}.

Based on the previous definition, we can define *reachability* in the context of abstract argumentation frameworks.

Definition 2.5 [Reachability] Let $AF = (AR, AT)$ be an argumentation framework. We say that given two arguments $a, b \in AR$, in AF, b is reachable from a iff there exists a path $P_{a,b}$ or $a = b$.

Roughly speaking, a strongly connected component in an argumentation framework $AF = (AR, AT)$ is a maximal set $S \subseteq AR$ such that every argument in S is reachable from ever other argument in S.

Definition 2.6 [Strongly Connected Components] Let $AF = (AR, AT)$ be an argumentation framework. $S \subseteq AR$ is a strongly connected component of AF iff $\forall a, b \in S$, a is reachable from b and b is reachable from a and $\nexists c \in AR \setminus S$, such that a is reachable from c and c is reachable from a. We denote the strongly connected components of AF by $SCCS(AF)$.

The concept of argumentation framework expansions describes the relationship between two argumentation frameworks.

Definition 2.7 [Argumentation Framework Expansions [6]] Let $AF = (AR, AT)$ and $AF' = (AR', AT')$ be argumentation frameworks. AF' is an *expansion* of AF (denoted by $AF \preceq_E AF'$) iff $AR \subseteq AR'$ and $AT \subseteq AT'$. AF' is a *normal expansion* of AF (denoted by $AF \preceq_N AF'$) iff $AF \preceq_E AF'$ and $(AR \times AR) \cap (AT' \setminus AT) = \{\}$.

An argumentation *semantics* σ is a function that takes an argumentation framework $AF = (AR, AT)$ as its input and returns a set of *extensions* $ES \subseteq 2^{AR}$. We say that a semantics σ is *universally defined* iff for every argumentation framework AF, it holds true that $|\sigma(AF)| \geq 1$. Let us introduce the argumentation semantics that are relevant in the context of our work, starting with some of the semantics that Dung introduces in his seminal paper.

Definition 2.8 [Complete and Grounded Semantics [8]] Let $AF = (AR, AT)$ be an argumentation framework. An admissible set $S \subseteq AR$ is:

- a *complete extension* iff each argument that is defended by S belongs to S. $\sigma_{complete}(AF)$ denotes the complete extensions of AF.

- a *grounded extension* of AF iff S is the minimal (w.r.t. set inclusion) complete extension of AF. $\sigma_{grounded}(AF)$ denotes the grounded extensions of AF.

Let us note that every argumentation framework has exactly one grounded extension [5]. Dung's semantics are based on the notion of an admissible set, while some other semantics are based on maximal conflict-free (or: *naive*) sets.

Definition 2.9 [Naive and Stage Semantics [13]] Let $AF = (AR, AT)$ be an argumentation framework and let $S \subseteq AR$.

- S is a naive extension of AF iff S is a maximal conflict-free subset of AR w.r.t. set inclusion. Naive semantics $\sigma_{naive}(AF)$ denotes all naive extensions of AF.
- S is a stage extension of AF iff $S \cup S^+$ is maximal w.r.t. set inclusion among all conflict-free sets, *i.e.*, there is no conflict-free set $S' \subseteq AR$, such that $(S' \cup S'^+) \supset (S \cup S^+)$. $\sigma_{stage}(AF)$ denotes all stage extensions of AF.

Some naive-based argumentation semantics are defined using an SCC-recursive approach. Before introducing these semantics, let us provide the definitions of a simple preliminary. For the sake of conciseness, we do not explain the SCC-recursive approach in detail; for this, let us refer to the corresponding paper by Baroni *et al.* [4].

Definition 2.10 [UP [4]] Let $AF = (AR, AT)$ be an argumentation framework. Let $E \subseteq AR$ and let S be a strongly connected component of AF ($S \in SCCS(AF)$). We define $UP_{AF}(S, E) = \{a \in S | \nexists b \in E \setminus S \text{ such that } (b, a) \in AT\}$.

Let us now introduce the definition of the SCC-recursive CF2 and stage2 semantics.

Definition 2.11 [CF2 and stage2 Semantics [4,9]] Let $AF = (AR, AT)$ be an argumentation framework and let $E \subseteq AR$. E is a CF2 extension iff:

- E is a naive extension of AF if $|SCCS(AF)| = 1$;
- $\forall S \in SCCS(AF)$, $(E \cap S)$ is a CF2 extension of $AF \downarrow_{UP_{AF}(S,E)}$, otherwise.

E is a stage2 extension iff:

- E is a stage extension of AF if $|SCCS(AF)| = 1$;
- $\forall S \in SCCS(AF)$, $(E \cap S)$ is a stage2 extension of $AF \downarrow_{UP_{AF}(S,E)}$, otherwise.

Argumentation principles that are relevant in the context of this paper are naivety and directionality. Let us first introduce the definition of naivety.

Definition 2.12 [Naivety [3]] Let σ be an argumentation semantics. σ satisfies the naivety principle iff for every argumentation framework $AF = (AR, AT)$ it holds true that $\forall E \in \sigma(AF)$, $E \in \sigma_{naive}(AF)$.

The directionality principle depends on the notion of an *unattacked set*.

Definition 2.13 [Unattacked Sets [3]] Let $AF = (AR, AT)$ be an argumentation framework. A set $S \subseteq AR$ is *unattacked* iff $\nexists a \in AR \setminus S$ such that a attacks S. $US(AF)$ denotes all unattacked sets in AF.

Let us provide the definition of the directionality principle.

Definition 2.14 [Directionality [3]] An argumentation semantics σ is directional iff for every argumentation framework $AF = (AR, AT)$, for every unattacked set of arguments $U \subseteq AR$ it holds true that $\sigma(AF \downarrow_U) = \{E \cap U | E \in \sigma(AF)\}$.

Finally, let us introduce the notion of an argumentation framework sequence (which is similar to notions of *expansion chains* as introduced by Baumann and Brewka [6]).

Definition 2.15 [Argumentation Framework Sequence] An argumentation framework sequence is a sequence $AFS = \langle AF_0, ..., AF_n \rangle$ where every $AF_i, 0 \le i \le n$ is an argumentation framework. An argumentation seuquence AFS is normally expanding iff it holds true for every $AF_j, 0 \le j \le n-1$ that $AF_j \preceq_N AF_{j+1}$.

3 Sequential Shkop Semantics

Before we formally introduce Shkop semantics, let us provide an intuition of the underlying approach. We assume we construct an argumentation framework argument-by-argument in an iterative manner and determine exactly one extension at each iteration step. We start with the empty set as our extension E_{-1}. At each step i, $0 \le i \le n$ (where $n+1$ is the number of arguments we add to our framework), we proceed as follows. We *test* E_{i-1}, given the current argumentation framework. From an intuitive perspective, we can colloquialise this test using the following question:

> *Given the current argumentation framework, can we without reasonable doubt keep the previous inference result?*

If we cannot answer this question in the affirmative, the test fails and E_i, as well as all following extensions are annotated as $false$, which indicates failure of meaningful inference. This – in turn – indicates that we need to re-arrange the order of arguments in our sequence so that the approach allows for meaningful inference. If the test passes, we define $E_i = E_{i-1} \cup \{a\}$, where a is our newly added argument, if $E_{i-1} \cup \{a\}$ is conflict-free; otherwise, we define $E_i = E_{i-1}$.

Let us formalise this approach. Based on the definition of an argumentation sequence, we can define the notion of a Shkop sequence.

Definition 3.1 [Shkop Sequence] Let $AFS = \langle AF_0, ..., AF_n \rangle$ be an argumentation framework sequence. AFS is a Shkop sequence iff:

- $AF_0 = (AR_0, AT_0)$, such that $|AR_0| = 1$;

- AFS is normally expanding;

- For every $AF_i, 0 \le i < n$, such that $AF_i = (AR_i, AT_i)$, it holds true that $|AR_{i+1} \setminus AR_i| = 1$.

Given an argumentation framework $AF = (AR, AT)$, a Shkop sequence $AFS = \langle AF_0, ..., AF_n \rangle = \langle (AR_0, AT_0), ..., (AR_n, AT_n) \rangle$ is a Shkop sequence of AF iff $AF_0 = AF \downarrow_{\{a\}}, a \in AR, AF_n = AF$, and for every $AF_i = (AR_i, AT_i), 1 \le i < n$ it holds true that $AR_i \subset AR$ and $AF_i = AF \downarrow_{AR_i}$.

Let us now define the notion of a *Shkop test*.

Definition 3.2 [Shkop Test] A Shkop test is a boolean function f that takes an argumentation framework $AF = (AR, AT)$, a set $S \subseteq AR$, and an argument $a \subseteq AR$ as its input.

We formalise the *basic Shkop* approach for determining the extensions of Shkop sequences, analogously to the intuition provided above.

Definition 3.3 [Basic Shkop] Given a Shkop sequence $AFS_{Shkop} = \langle AF_0, ..., AF_n \rangle = \langle (AR_0, AT_0), ..., (AR_n, AT_n) \rangle$, the basic Shkop function $s_{Shkop,f}$ returns a sequence of tuples $\langle T_0, ..., T_n \rangle = \langle (E_0, t_0), ..., (E_n, t_n) \rangle$ such that for $0 \leq i \leq n$, $E \subseteq AR_i$, $t_i \in \{true, false\}$ and:

$T_i = (\{a' | a' \in AR_0, (a', a') \notin AT_0\}, true)$ if $i = 0$;

$T_i = (\{\}, false)$ else if t_{i-1} is false or

$(\exists b \in nsa(AR_{i-1})$, such that in $nsa(AF_i)$, b is reachable from a

and a is not reachable from b) or

$f(nsa(AF_i), E_{i-1}, a)$ is false;

$T_i = (E_{i-1} \cup \{a\}, true)$ else if t_{i-1} is true and $E_{i-1} \cup \{a\}$ is conflict-free;

$T_i = T_{i-1}$, otherwise,

where a is the only argument in $AR_i \setminus AR_{i-1}$ and f is a Shkop test.

The above definition has the obvious problem that we may infer a tuple $(\{\}, false)$ from an argumentation framework, which indicates that we do not have a meaningful inference result. This comes in handy when we define our approach for abstract argumentation semantics design, but requires adjustments to allow for a better sequential perspective.

Definition 3.4 [Sequential Shkop] Given a Shkop sequence $AFS_{Shkop} = \langle AF_0, ..., AF_n \rangle = \langle (AR_0, AT_0), ..., (AR_n, AT_n) \rangle$, the sequential Shkop function $seq_{Shkop,f}$ returns a sequence of sets of arguments $\langle E_0, ..., E_n \rangle$ such that for each $E_i, 0 \leq i \leq n$:

$E_i = \{a' | a' \in AR_0, (a', a') \notin AT_0\}$ if $i = 0$;

$E_i = E_i^i$ else if $(\exists b \in nsa(AR_{i-1})$, such that in $nsa(AF_i)$, b is reachable from a

and a is not reachable from b) or

$f(nsa(AF_i), E_{i-1}, a)$ is false;

$E_i = E_{i-1} \cup \{a\}$ else if $E_{i-1} \cup \{a\}$ is conflict-free;

$E_i = E_{i-1}$, otherwise,

where:

- a is the only argument in $AR_i \setminus AR_{i-1}$;
- $\langle E_0^i, ..., E_n^i \rangle = seq_{Shkop,f}(AFS_{Shkop}^i)$, such that $AFS_{Shkop}^i = \langle AF_0^i, ..., AF_n^i \rangle = \langle (AR_0^i, AT_0^i), ..., (AR_n^i, AT_n^i) \rangle$ and for $0 \leq j \leq n, j \neq i-1$, $AF_j^i = AF_i$ and $AF_{i-1}^i = AF_n \downarrow_{AR_{i-2}^i \cup \{a\}}$ if $i-2 \geq 0$, $AF_{i-1}^i = AF_n \downarrow_{\{a\}}$ otherwise;
- f is a Shkop test.

51

Roughly speaking, when expanding an argumentation framework one argument at a time, if we encounter an inference result that fails our Shkop test or if our new argument is a not reachable, "upstream" attacker of existing arguments, we re-arrange our Shkop sequence so that the newly added argument switches places in the sequence with its predecessor. Let us informally claim that this never causes an infinite loop if we use the grounded Shkop test as defined below.

Definition 3.5 [Grounded Shkop Test] Let $AF = (AR, AT)$ be an argumentation framework and $a \in AT$. The grounded Shkop test g_{Shkop} is a Shkop test that takes AF, $S \subseteq AR$, and $a \in AR$ as its inputs and generates its output as follows:

$$g_{Shkop}(AF, S, a) = \begin{cases} true & \text{if } S \cup \sigma_{grounded}(AF \downarrow_{AR'}) \text{ is conflict-free;} \\ false & \text{otherwise,} \end{cases}$$

where $AR' = \{b | b \in AR, b \in S \text{ or } b \text{ is reachable from } a\}$. The basic Shkop function that applies the grounded Shkop test is denoted by s_{SG}, and the sequential Shkop function that applies the grounded Shkop test is denoted by seq_{SG}.

Let us illustrate the approach by example.

Example 3.6 Consider an agent who uses the sequential Shkop approach with the grounded Shkop test to decide whether to take an umbrella when leaving the house or not. The agent creates an initial argumentation framework $AF_0 = (\{u\}, \{\})$, where the argument u denotes the action of taking the umbrella. Obviously, AF_0 is resolved as $\{u\}$. Then, the agent looks out of the window and sees that it is not raining (denoted by $\neg r$), and constructs the next argumentation framework $AF_1 = (\{u, \neg r\}, \{(\neg r, u)\})$. $\neg r$ can reach u, but not vice versa); also, $\{\neg r\}$ is the grounded extension of $nsa(AF_1)$ and in conflict with $\{u\}$. Hence, the agent cannot reach any conclusion given the current Shkop sequence. Our agent re-arranges the sequence, and now starts with $AF_0' = (\{\neg r\}, \{\})$, which she resolves as $\{\neg r\}$. Then, she expands AF_0' to AF_1. Because in AF_1, $\neg r$ is not reachable from u and also not attacked by the grounded extension of $nsa(AF_1)$, the agent concludes, according to sequential Shkop, $\{\neg r\}$. While already out in the street, the agent checks the weather forecast on the mobile phone. It indicates that it could start raining soon (r), i.e., the agent constructs the argumentation framework $AF_2 = (\{u, \neg r, r\}, \{(\neg r, u), (\neg r, r), (r, \neg r)\})$[5]. Because u is reachable from r, but r is not reachable from u, our agent again needs to re-arrange the Shkop sequence and obtains $AFS = \langle AF_0', (\{\neg r, r\}, \{(\neg r, r), (r, \neg r)\}), AF_2 \rangle$. From AF_0', she has already inferred $\neg r$. When expanding AF_0' to $(\{\neg r, r\}, \{(\neg r, r), (r, \neg r)\})$, the new argument r is in conflict with the current conclusion $\{\neg r\}$; but because $\{\neg r\}$ is not attacked by the grounded extension of $nsa(AF_2)$, r is discarded. Similarly, u is discarded after the expansion to AF_2. This means that again, our agent remains with the conclusion $\{\neg r\}$. Figure 1 depicts the argumentation frameworks of the example.

---

[5] We assume the agent does not have any preferences w.r.t. the following scenarios: 1) having an umbrella with her although it is not raining; 2) having no umbrella with her although it is raining.

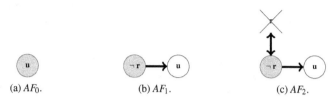

(a) AF_0.   (b) AF_1.   (c) AF_2.

Fig. 1. $AF = (\{u, r, \neg r\}, \{(r, \neg r), (\neg r, r), (\neg r, u)\})$.

4   (Grounded) Shkop Semantics

Let us extend the Shkop approach so that it can determine the extension of an argumentation framework without requiring a Shkop sequence.

Definition 4.1 [Shkop Semantics] Let $AF = (AR, AT)$ be an argumentation framework, let f be a Shkop test and let $s_{Shkop,f}$ be a basic Shkop function. Let $SEQS = \{AFS_0, ..., AFS_n\}$ be the set of all Shkop sequences of AF. If $AR = \{\}$, $\{\}$ is the only Shkop$_f$ extension of AF. Otherwise, $S \subseteq AR$ is a Shkop$_f$ extension of AF iff $\exists AFS \in SEQS$, such that $s_{Shkop,f}(AFS) = \langle(E_0, true), ..., (E_m, true)\rangle$ and $S = E_m$. $\sigma_{Shkop,f}(AF)$ denotes all Shkop$_f$ extensions of AF.

Note that the definition of a particular Shkop semantics requires the specification of a Shkop test. The Shkop semantics we analyse in this paper uses the grounded Shkop test (Definition 3.5).

Definition 4.2 [Grounded Shkop Semantics] Let $AF = (AR, AT)$ be an argumentation framework and let g_{Shkop} be the grounded Shkop test. $S \subseteq AR$ is a grounded Shkop extension iff $S \in \sigma_{Shkop,g_{Shkop}}(AF)$. The shorthand notation $\sigma_{SG}(AF)$ denotes all grounded Shkop extensions of AF.

Let us illustrate by example how grounded Shkop semantics works.

Example 4.3 Let us again consider the scenario presented by Example 3.6. However, now we do not assume that the argumentation framework is constructed and resolved iteratively, *i.e.*, we start wit $AF = (AR, AT) = (\{u, \neg r, r\}, \{(\neg r, u), (\neg r, r), (r, \neg r)\})$. We first generate all permutation sequences of AR that satisfy the "reachability constraint", *i.e.* the order of a permutation sequence respects the partial order that is established by the acyclic directed SCC graph (roughly speaking):

$$PERS_{AR} = \left\{ \begin{array}{l} \langle\neg r, r, u\rangle, \\ \langle r, \neg r, u\rangle \end{array} \right\}$$

Then, we generate the Shkop sequences based on the permutations:

$$SEQS = \left\{ \begin{array}{ll} \langle(\{\neg r\}, \{\}), & (\{\neg r, r\}, \{(\neg r, r), (r, \neg r)\}), & (\{u, \neg r, r\}, \{(\neg r, u), (\neg r, r), (r, \neg r)\})\rangle, \\ \langle(\{r\}, \{\}), & (\{\neg r, r\}, \{(\neg r, r), (r, \neg r)\}), & (\{u, \neg r, r\}, \{(\neg r, u), (\neg r, r), (r, \neg r)\})\rangle \end{array} \right\}$$

We resolve the Shkop sequences using the grounded Shkop test:

$$\{s_{SG}(AFS)|AFS \in SEQS\} = \left\{ \begin{array}{lll} \langle(\{\neg r\}, true), & (\{\neg r\}, true), & (\{\neg r\}, true)\rangle, \\ \langle(\{r\}, true), & (\{r\}, true), & (\{u, r\}, true)\rangle \end{array} \right\}$$

53

Now, we take, for each Shkop sequence, the last extension the basic Shkop approach has determined (if not annotated as $false$)[6], which gives us $\sigma_{SG}(AF) = \{\{u,r\},\{\neg r\}\}$.

Another step-by-step example of how Shkop semantics works is provided in Appendix A, and a potential application scenario is outlined in Appendix B; more examples are available as test specifications of the implementation (https://git.io/J07e5).

5 Analysis

In this section, we show (by principle-based analysis) that grounded Shkop is a universally defined and directional naive set-based semantics and (by using examples) that it has some advantages in comparison to some other naive set-based semantics. Let us start by making some straightforward observations.

Proposition 5.1 *For every argumentation framework* $AF = (AR, AT)$, *let* $E'_{grounded}$ *be the grounded extension of* $nsa(AF)$. $\forall E \in \sigma_{SG}(AF)$, *it holds true that* $E'_{grounded} \subseteq E$.

Note that the proofs of all propositions are provided in Appendix C.

Arguments that are attacked by the grounded extension are not entailed by any grounded Shkop extension.

Proposition 5.2 *For every argumentation framework* $AF = (AR, AT)$, *let* $E'_{grounded}$ *be the grounded extension of* $nsa(AF)$. $\forall E \in \sigma_{SG}(AF)$, *it holds true that* $E \cap E^+_{grounded} = \{\}$.

We formally observe that grounded Shkop semantics is universally defined.

Proposition 5.3 σ_{SG} *is universally defined.*

Grounded Shkop semantics satisfies the naivety principle, *i.e.* given an argumentation framework $AF = (AR, AT)$, every grounded Shkop extension of AF is a $\subseteq$-maximal conflict-free subset of AR.

Proposition 5.4 σ_{SG} *satisfies the naivety principle.*

Also, grounded Shkop semantics satisfies directionality.

Proposition 5.5 σ_{SG} *satisfies the directionality principle.*

Let us introduce two examples that illustrate well-known counter-intuitive behaviour of some naive set-based semantics, in particular CF2 and stage2 semantics.

Example 5.1 Let us have the argumentation framework $AF = (\{a,b,c,d,e,f\}, \{(a,b),(b,c),(c,d),(d,e),(e,f),(f,a)\})$. Because AF consists of a single strongly connected component, we have $\sigma_{CF2}(AF) = \sigma_{naive}(AF) = \{\{a,c,e\},\{b,d,f\},\{a,d\},\{b,e\},\{c,f\}\}$. The extensions that contain exactly two arguments are counter-intuitive. Let us take $\{a,d\}$. Assuming a is in the extension,

---

[6] Let us note that the Shkop test indeed fails in some scenarios in which the "reachability constraints" are respected, consider the argumentation framework $AF = (\{a,b,c,d\}, \{(a,c),(b,c),(c,d),(d,b)\})$ and a Shkop sequence that reflects the alphabetical order of the arguments.

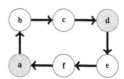

Fig. 2. $AF = (\{a,b,c,d,e,f\}, \{(a,b),(b,c),(c,d),(d,e),(e,f),(f,a)\})$. $\{a,d\} \in \sigma_{CF2}(AF)$. This is counter-intuitive because there is an "uninterrupted" indirect attack from a to d.

b is successfully attacked, and hence c should be in the extension, successfully attacking d. Let us note that $\sigma_{SG}(AF) = \sigma_{stage}(AF) = \{\{a,c,e\}, \{b,d,f\}\}$. Figure 2 depicts AF.

Cramer and Van der Torre provide a principle-based analysis of this issue with their Strong Completeness Outside Odd Cycles (SCOOC) principle [7]; analysing grounded Shkop semantics using this principle can be considered relevant future work. Still, let us highlight that in the way grounded Shkop solves the well-known problem of CF2 semantics with even-length cycles with six or more arguments lies a key difference between the design approaches of grounded Shkop semantics, stage2 semantics, and SCF2 semantics. stage2 semantics solves the problem by applying a more sceptical semantics (stage semantics) to determine the extensions of SCCS, which leads to semantics behaviour that is more sceptical in many other aspects as well (consider, for example $AF = (\{a,b,c\}, \{(a,b),(b,a),(b,c),(c,a)\})$). SCF2 semantics addresses the issue by defining a principle that "catches" the even-cycle problem and by enforcing this principle explicitly; hence, let us claim that the general approach of the semantics lacks a well-motivated intuition. In contrast, the problem does not need any explicit fix in grounded Shkop semantics.

Another example illustrates how grounded Shkop semantics handles self-attacking arguments, which, in contrast to the even-length cycle problem, requires an explicit (but rather straightforward) fix.

Example 5.2 Let us have the argumentation framework $AF' = (\{a,b,c,d,e\}, \{(a,b),(b,c),(c,d),(d,e),(e,a),(a,a),(d,d)\})$. Because AF' consists of only one strongly connected component, we have $\sigma_{stage2}(AF') = \sigma_{stage}(AF') = \{\{b,e\}, \{c,e\}\}$. However, let us note that when we "discard" the self-attacking arguments, we get $AF = (\{b,c,e\}, \{(b,c)\})$ and $\sigma_{stage2}(AF') = \{\{b,e\}\}$; i.e., the addition of self-attacking arguments changes the conclusions σ_{stage2} infers, arguably in a counter-intuitive manner. Cramer and Van der Torre provide a principle-based analysis of this issue by introducing the Irrelevance of Necessarily Rejected Arguments (INRA) principle [7]. Let us note that $\sigma_{SG}(AF') = \{\{b,e\}\}$. Figure 3 depicts AF'.

Let us formally observe that grounded Shkop semantics ignores self-attacking arguments.

Proposition 5.6 *For every argumentation framework AF, it holds true that* $\sigma_{SG}(AF) = \sigma_{SG}(nsa(AF))$.

In addition, in Appendix D we provide the proof that grounded Shkop semantics satisfies the recently introduced *weak reference independence* principle [10] that is

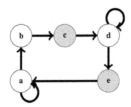

Fig.   3. $AF' = (\{a,b,c,d,e\}, \{(a,b),(b,c),(c,d),(d,e),(e,a),(a,a),(d,d)\})$.
Given stage semantics σ_{stage} and stage2 semantics σ_{stage2} , it holds true that
$\{c,e\} \in \sigma_{stage}(AF'), \{c,e\} \in \sigma_{stage2}(AF')$. This is counter-intuitive because c is attacked by b,
which is only attacked by the self-attacking argument a.

based on the notion of consistent preferences of a rational decision-maker in microe-
conomic theory.

6   Conclusion

We have introduced the Shkop approach to sequential argumentation. The approach
allows us to define naive set-based argumentation semantics that construct extensions
on an argument-by-argument basis, in contrast to existing approaches that are often
recursive on strongly connected component-level. The newly introduced grounded
Shkop argumentation semantics has advantages over some other naive set-based argumentation semantics, in particular over CF2 semantics. Open issues that future research may address remain, for example an analysis that formally compares grounded
Shkop semantics and SCF2 semantics [7], as well as a more comprehensive principle-
based analysis of grounded Shkop semantics.

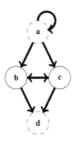

Fig. 4. $AF = (AR,AT) = (\{a,b,c,d\},\{(a,a),(a,b),(a,c),(b,c),(c,b),(b,d),(c,d)\})$.

Appendix

Appendix A - Another Step-by-Step Shkop Example

Let us provide another step-by-step example that illustrates how grounded Shkop semantics works.

Example 6.1 Let us have the argumentation framework $AF = (AR,AT) = (\{a,b,c,d\},\{(a,a),(a,b),(a,c),(b,c),(c,b),(b,d),(c,d)\})$ (Figure 4). We generate all permutations of AR that satisfy the "reachability constraint" (roughly speaking), but place a always first because we know its place in any sequence is irrelevant (not that the reachability constraint ignores self-attacking arguments); i.e., we get:

$$PERS'_{AR} = \left\{ \begin{array}{l} \langle a,b,c,d \rangle, \\ \langle a,c,b,d \rangle \end{array} \right\}$$

Then, we generate the Shkop sequences based on the permutations:

$$SEQS = \left\{ \begin{array}{llll} \langle AF \downarrow \{a\}, & AF \downarrow \{a,b\}, & AF \downarrow \{a,b,c\}, & AF \rangle, \\ \langle AF \downarrow \{a\}, & AF \downarrow \{a,c\}, & AF \downarrow \{a,b,c\}, & AF \rangle \end{array} \right\}$$

We resolve the Shkop sequences using the grounded Shkop test:

$$\{s_{SG}(AFS)|AFS \in SEQS\} = \left\{ \begin{array}{llll} \langle(\{\},true), & (\{b\},true), & (\{b\},true), & (\{b\},true)\rangle, \\ \langle(\{\},true), & (\{c\},true), & (\{c\},true), & (\{c\},true)\rangle \end{array} \right\}$$

Now, we take, for each Shkop sequence, the last extension the basic Shkop approach has determined (if not annotated as *false*), which gives us $\sigma_{SG}(AF) = \{\{b\},\{c\}\}$.

Appendix B - Shkop Application Examples

Let us consider an application scenario for the Shkop approach.

Example 6.2 We have a software application landscape where several reasoning engines draw inferences from heterogeneous knowledge sources. The inference results of different engines may be conflicting, and need to be combined to draw aggregated conclusions that inform tactical decision-making. We start with a statement a that has been inferred by engine $\mathscr{R}_1$: $AF_0 = (\{a\},\{\})$. Initially, no conflicting inferences have been drawn, and the organisation bases its decisions on $\{a\}$. After a while, engine $\mathscr{R}_2$ infers statement b, and a and b attack each other: $AF_1 = (\{a,b\},\{(a,b),(b,a)\})$. Still,

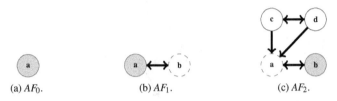

(a) AF_0.   (b) AF_1.   (c) AF_2.

Fig. 5. $AF_2 = (\{a,b,c,d\}, \{(a,b),(b,a),(c,a),(c,d),(d,a),(d,c)\})$.

we do not want our organisation to change course, considering that we need to keep a steady course in an ever-changing environment. $\{a\}$ passes the grounded Shkop test, and we keep our initial inference result $\{a\}$. Figure 5 depicts the argumentation frameworks of the example. However, at a later point in time, both $\mathscr{R}_1$ and $\mathscr{R}_2$ update their inferences, and provide statements that c and d that attack a, but also attack each other: $AF_2 = (\{a,b,c,d\}, \{(a,b),(b,a),(c,a),(c,d),(d,a),(d,c)\})$. We have not established a total order on the set of arguments and do not know whether we should infer c or d, but no matter the order, we know that we have to infer "either c or d"; i.e., we have compelling evidence that we should no longer infer a, even if this implies change efforts for our organisation.

Appendix C - Proofs

This appendix contains the proofs of the propositions provided in the analysis section. Propositions are re-stated and have the same numbering as the corresponding propositions in Section 5.

Proposition 5.1   *For every argumentation framework $AF = (AR, AT)$, let $E'_{grounded}$ be the grounded extension of $nsa(AF)$. $\forall E \in \sigma_{SG}(AF)$, it holds true that $E'_{grounded} \subseteq E$.*

Proof.

(i) By definition of s_{SG}, for every Shkop sequence $AFS = \langle (AR_0, AT_0), ..., (AR_n, AT_n) \rangle$ of AF, for $s_{SG}(AFS) = \langle (E_0, t_0), ..., (E_n, t_n) \rangle$, for every $E_i, t_i, 0 < i \leq n$ the following statement holds true given a as the only argument in $AR_i \setminus AR_{i-1}$:

If t_i is true then

(in $AF_i, \forall b \in AR_{i-1}$ if b is reachable from a then a is reachable from b) and

$g_{Shkop}(nsa(AF), E_{i-1}, a)$

(ii) From *i)* it follows that by definition of s_{SG} (which entails the definition of the grounded Shkop test g_{Shkop}), for every Shkop sequence $AFS = \langle (AR_0, AT_0), ..., (AR_n, AT_n) \rangle$ of AF, for $s_{SG}(AFS) = \langle (E_0, t_0), ..., (E_n, t_n) \rangle$, for every $E_i, t_i, 0 \leq i \leq n$ if t_i is true then the following statement holds true:

$$E' \subseteq E_i, E' \in \sigma_{grounded}(nsa((AR_i, AT_i)))$$

(iii) From   *ii)*   it   follows   that   for   every   Shkop   sequence $AFS =$ $\langle (AR_0, AT_0), ..., (AR_n, AT_n) \rangle$ of AF,   for $s_{SG}(AFS) = \langle E_0, ..., E_n \rangle$,   if t_n is true for E_n the following statement holds true:

$$E' \subseteq E_n, E' \in \sigma_{grounded}(nsa((AR, AT)))$$

(iv) From *iii)* it follows that by definition of $\sigma_{SG}, \forall E \in \sigma_{SG}(AF)$, it holds true that $E'_{grounded} \subseteq E$. This proves the proposition.

$\square$

Proposition 5.2 *For every argumentation framework $AF = (AR, AT)$, let $E'_{grounded}$ be the grounded extension of $nsa(AF)$. $\forall E \in \sigma_{SG}(AF)$, it holds true that $E \cap E^+_{grounded} = \{\}$.*

Proof. Because by definition of σ_{SG}, for every argumentation framework AF, each σ_{SG}-extension of AF is conflict-free, the proof follows directly from Proposition 5.1.$\square$

Proposition 5.3 σ_{SG} *is universally defined.*

Proof.

(i) Let us observe that if $AF = (\{\}, \{\})$, then $\sigma_{SG} = \{\{\}\}$ (by definition). By definition of a Shkop sequence (Definition 3.1), and considering that every non-empty argumentation framework is a directed graph and hence its strongly connected components form a directed acyclic graph, for every non-empty argumentation framework AF, there exists a Shkop sequence AFS of AF, such that $AFS = \langle AF_0, ..., AF_n \rangle$ and the following statement[7] holds true for $AF_i = (AR_i, AT_i), 0 \le i < n$ and given a, such that a is the only argument in $AR_{i+1} \setminus AR_i$:

$$\forall b \in AR_i \text{ if } b \text{ is reachable from } a \text{ in } nsa(AF_{i+1}) \text{ then}$$
$$a \text{ is reachable from } b \text{ in } nsa(AF_{i+1})$$

(ii) From *i)* it follows that by definition of grounded semantics, for every non-empty argumentation framework AF, there exists a Shkop sequence AFS of AF, such that $AFS = \langle AF_0, ..., AF_n \rangle$, $s_{SG}(AFS) = \langle (E_0, true)..., (E_n, t) \rangle$ and the following statement[8] holds true for $AF_i = (AR_i, AT_i), 0 \le i < n$ and given a as only argument in $AR_{i+1} \setminus AR_i$ and $AR' = \{b | b \in AR_{i+1}, b \in E_i$ or b is reachable from $a\}$:

$$\text{if } a \text{ attacks } AR_i \text{ then } \sigma_{grounded}(nsa(AF_{i+1} \downarrow_{AR'})) \subseteq \sigma_{grounded}(nsa(AF'_i)),$$

where $AF'_i = AF_i$ if $i = 0$; $AF'_i = AF_i \downarrow_{AR'_i}$, $AR'_i = \{b | b \in AR_i, b \in E_{i-1}$ or b is reachable from $a'\}$ and a' is the only argument in $AR_i \setminus AR_{i-1}$, otherwise.

(iii) From *ii)* it follows that by definition of basic Shkop (Definition 3.3), it holds true that there exists a Shkop sequence $AFS = \langle AF_0, ..., AF_n \rangle$ of AF, such that $s_{Shkop}(AFS) = \langle (E_0, t_0), ..., (E_n, t_n) \rangle$ and for k, $0 \le k \le n$, t_k is true and hence t_n is true.

---

[7] In words: we can construct a Shkop sequence of AF such that an argument a attacks an argument it succeeds in the sequence (directly or indirectly) if and only if a closes a loop with this argument (ignoring self-attacking arguments).

[8] If we have a loop (ignoring self-attacking arguments) – and hence the sequence cannot be expanded following a clear attack direction – we can select the argument that closes the loop such that this argument is not in the grounded extension of the current argumentation framework's restriction that excludes self-attacking arguments and arguments that are not reachable from the newly added argument and in conflict with the previous inference result.

(iv) From *iii)* it follows that by definition of grounded Shkop semantics (Definition 4.2), for every argumentation framework AF it holds true that $|\sigma_{SG}(AF)| \geq 1$. This proves the proposition. $\qquad\square$

Proposition 5.4 σ_{SG} *satisfies the naivety principle.*

Proof. $\sigma_{SG}(AF)$ satisfies the naivety principle (Definition 2.12) iff for every argumentation framework $AF = (AR, AT)$ it holds true that $\forall E \in \sigma_{SG}(AF)$, E is a $\subseteq$-maximal conflict-free set of AF. By definition of s_{SG}, for every argumentation framework $AF = (AR, AT)$, for every Shkop sequence $AFS = \langle AF_0, ..., AF_n \rangle$ of AF, $s_{SG}(AFS) = \langle (E_0, t_0), ..., (E_n, t_n) \rangle$, for $0 \leq i \leq n$, t_i is false or E_i is a $\subseteq$-maximal conflict-free set of AF_i. It follows that by definition of σ_{SG}, $\forall E \in \sigma_{SG}$, E is a $\subseteq$-maximal conflict-free set of AF. This proves the proposition. $\qquad\square$

Proposition 5.5 σ_{SG} *satisfies the directionality principle.*

Proof.

(i) σ_{SG} satisfies the directionality principle (Definition 2.14) iff for every argumentation framework $AF = (AR, AT)$, for every unattacked set of arguments $U \subseteq AR$ it holds true that $\sigma_{SG}(AF \downarrow_U) = \{E \cap U | E \in \sigma_{SG}(AF)\}$.

(ii) By definition of s_{SG} (Definition 3.3), for every argumentation framework $AF = (AR, AT)$ and for every Shkop sequence $AFS = \langle AF_0, ..., AF_n \rangle = \langle (AR_0, AT_0), ..., (AR_n, AT_n) \rangle$ of AF and $s_{SG}(AFS) = \langle E_0, ..., E_n \rangle$, for $0 \leq i < n, i \leq j \leq n$ it holds true that if $AR_j \setminus AR_i$ does not attack AR_i then $E_i = E_j \cap AR_i$.

(iii) From 2. it follows that by definition of σ_{SG} (Definition 4.2), for every unattacked set of arguments $U \subseteq AR$ it holds true that $\sigma_{SG}(AF \downarrow_U) = \{E \cap U | E \in \sigma_{SG}(AF)\}$. This proves the proposition.

$\qquad\square$

Proposition 5.6 *For every argumentation framework AF, it holds true that* $\sigma_{SG}(AF) = \sigma_{SG}(nsa(AF))$.

Proof. For every non-empty argumentation framework AF, for every Shkop sequence $AFS = \langle (AR_0, AT_0), ..., (AR_n, AT_n) \rangle$ of AF, for $s_{SG}(AFS) = \langle (E_0, t_0), ..., (E_n, t_n) \rangle$ and $i, 0 \leq i \leq n$, we have the following cases:

Case 1: $i = 0$. By definition of s_{SG} (Definition 3.3), for a, such that a is the only argument in AR_0 it holds true that if $(a, a) \in AT_i$ then $E_i = \{\}$ and t_i is true.

Case 2: $i > 0$. Let a be the only argument in $AR_i \setminus AR_{i-1}$. By definition of s_{SG} (Definition 3.3), it holds true that if $(a, a) \in AT_i$ then t_i is false iff t_{i-1} is false and if t_i is true then $E_i = E_{i-1}$.

From these two cases it follows that by definition of σ_{SG} (Definition 4.2), it holds true that $\sigma_{SG}(AF) = \sigma_{SG}(nsa(AF))$. This proves the proposition. $\qquad\square$

Appendix D - Grounded Shkop Semantics and Weak Reference Independence

The recently introduced weak reference independence principle assesses the consistency of an argumentation semantics by considering the extensions the semantics returns in a normal expansion process.

Definition 6.3 [Weak Reference Independence [10]] Let σ be an argumentation semantics. σ satisfies weak reference independence iff for every two argumentation frameworks $AF = (AR, AT), AF' = (AR', AT')$ such that $AF \preceq_N AF'$, $\forall E \in \sigma(AF)$ it holds true that $\exists E' \in \sigma(AF')$, such that $E' \not\subseteq AR$ or $E' = E$.

Let us prove that grounded Shkop semantics satisfies weak reference independence.

Proposition 6.4 (Grounded Shkop Semantics Satisfies Weak Reference Independence)
σ_{SG} *satisfies the weak reference independence principle.*

Proof. By definition (Definition 6.3), σ_{SG} satisfies the weak reference independence principle iff for every two argumentation frameworks $AF = (AR, AT), AF' = (AR', AT')$, such that $AF \preceq_N AF', \forall E \in \sigma_{SG}(AF)$ it holds true that $\exists E' \in \sigma_{SG}(AF')$, such that $E' \not\subseteq AR \vee E' = E$. Let us observe that given any $AF = (AR, AT), AF' = (AR', AT')$, such that $AF \preceq_N AF'$ and $AF' \neq (\{\}, \{\})$, there exists a Shkop sequence $AFS = \langle AF_0, ..., AF_j \rangle = \langle (AR_0, AT_0), ..., (AR_j, AT_j) \rangle$ of AF, such that $AF = AF_i$, $0 \leq i \leq j$ and $AF' = AF'_j$ (by definition of a Shkop sequence) and for $0 < m \leq n$, given a_m as the only argument in $AR_m \setminus AR_{m-1}$, $\forall b \in AR_{m-1}$ it holds true that if b is reachable from a_m in AF_m then a_m is reachable from b in AF_m (because σ_{SG} is universally defined, this follows from the definition of σ_{SG}).

Let us consider $n = j - i$. If $n = 0$, it follows that $AF = AF'$ and the proposition holds true. For $1 \leq n \leq j$, and $E_{j-n} \in \sigma_{SG}(AF)$, let us provide a proof by induction on n.

Base case: $n = 1$. Let a be the only argument in $AR_j \setminus AR_{j-1}$ and let $AR'_j = \{b | b \in AR_j, b \in E$ or b is reachable from $a\}$. From the definition of σ_{SG} it follows that if $\nexists E_j \in \sigma_{SG}(AF')$, such that $E_j \cap AR_{j-1} = E_i$ then $(AR' \setminus AR_{j-1}) \subseteq E'_g$, such that $E'_g \in \sigma_{grounded}(nsa(AF' \downarrow_{AR'_j}))$ and $a \in E'_g$. Consequently, it holds true that $\exists E' \in \sigma_{SG}(AF')$, such that $E' \not\subseteq AR$. This proves the proposition for the base case.

Induction case: $n = k + 1$. From the base case it follows that if $\nexists E_j \in \sigma_{SG}(AF')$, such that $E_j \cap AR_{j-(k+1)} = E_i$ then $\exists AF_l, j - (k+1) < l \leq j$, such that $\exists E_l \in \sigma_{SG}(AF_l)$ and $E_l \not\subseteq AR$. In turn, it follows that given E_l, such that $E_l \not\subseteq AR$, if $\nexists E_j \in \sigma_{SG}(AF')$, such that $E_j \cap AR_l = E_l$ then $\exists E_j \in \sigma_{SG}(AF_j)$, such that $E_j \not\subseteq AR_l$. This proves the proposition for the induction case.

$\square$

References

[1] Abraham, M., D. M. Gabbay and U. J. Schild, *The handling of loops in talmudic logic, with application to odd and even loops in argumentation*, HOWARD-60: A Festschrift on the Occasion of Howard Barringer's 60th Birthday (2014).

[2] Baroni, P., M. Caminada and M. Giacomin, *Abstract argumentation frameworks and their semantics*, in: P. Baroni, D. Gabbay, G. Massimiliano and L. van der Torre, editors, *Handbook of Formal Argumentation. College Publications*, College Publications, 2018 pp. 159–236.

[3] Baroni, P. and M. Giacomin, *On principle-based evaluation of extension-based argumentation semantics*, Artificial Intelligence **171** (2007), pp. 675 – 700, argumentation in Artificial Intelligence. URL http://www.sciencedirect.com/science/article/pii/S0004370207000744

[4] Baroni, P., M. Giacomin and G. Guida, *Scc-recursiveness: a general schema for argumentation semantics*, Artificial Intelligence **168** (2005), pp. 162 – 210.

[5] Baumann, R., *On the nature of argumentation semantics: Existence and uniqueness, expressibility, and replaceability*, Journal of Applied Logics **4** (2017), pp. 2779–2886.

[6] Baumann, R. and G. Brewka, *Expanding argumentation frameworks: Enforcing and monotonicity results.*, COMMA **10** (2010), pp. 75–86.

[7] Cramer, M. and L. van der Torre, *Scf2-an argumentation semantics for rational human judgments on argument acceptability*, in: *Proceedings of the 8th Workshop on Dynamics of Knowledge and Belief (DKB-2019) and the 7th Workshop KI & Kognition (KIK-2019) co-located with 44nd German Conference on Artificial Intelligence (KI 2019), Kassel, Germany, September 23, 2019*, 2019, pp. 24–35.

[8] Dung, P. M., *On the acceptability of arguments and its fundamental role in nonmonotonic reasoning, logic programming and n-person games*, Artificial intelligence **77** (1995), pp. 321–357.

[9] Gaggl, S. A., "Solving argumentation frameworks using answer set programming," TU Wien (Diploma Thesis), 2009.

[10] Kampik, T. and J. C. Nieves, *Abstract argumentation and the rational man*, Journal of Logic and Computation (2021), exab003.
URL https://doi.org/10.1093/logcom/exab003

[11] Thimm, M., *Tweety: A comprehensive collection of java libraries for logical aspects of artificial intelligence and knowledge representation*, in: *Proceedings of the Fourteenth International Conference on Principles of Knowledge Representation and Reasoning*, KR'14 (2014), p. 528–537.

[12] van der Torre, L. and S. Vesic, *The principle-based approach to abstract argumentation semantics*, IfCoLog Journal of Logics and Their Applications **4** (2017).

[13] Verheij, B., *Two approaches to dialectical argumentation: admissible sets and argumentation stages*, Proc. NAIC **96** (1996), pp. 357–368.

Spaces of Argumentation and Their Interaction

Some elements of thought inspired by controversies and dispute in France during the Covid-19 crisis

Gabriella Pigozzi and Juliette Rouchier

Université Paris-Dauphine
PSL Research University
LAMSADE
75016 Paris, France

1 Introduction

During the Covid-19 pandemic, many public policy decisions had to be taken. These decisions were taking place in an unusual context and using a very novel list of policy actions – such as lockdown, curfew or school closures. Because of their novelty, deciders needed to justify such decisions. As a consequence, we witnessed a very rapid construction and circulation of arguments in public spaces. One particularity in the Covid-19 debate is that most arguments were labelled as "scientific": scientists were counselling governments around the globe and became a regular presence on the media.

In this paper, we focus on the interactions between scientists and the media. We are interested in moments when scientific arguments are judged by worlds other than science (such as journalists or network media). In the opposite direction, journalists have a massive influence on scientists, who sometimes adopt their final judgment, although the reasoning does not follow their own usual standards.

In March 2020 Professor Didier Raoult (head of IHU in Marseille) had learnt by a Chinese colleague that chloroquine could work on the SARS-CoV-2 virus in vitro. After a few weeks, the IHU announced that there were signs of reduction of viral load in individuals after hydroxychloroquine (HCQ) was given and that the effect was even more striking if azytromycine (Az) was added. The fact that results were preliminary was not accepted, and the results were qualified as fake, because there was no statistical significance in the comparison of series. The debate on the HCQ in France became quickly very heated and was under the international spotlight when the President of the United States, Donald Trump, and the President of Brazil, Jair Bolsonaro, promoted HCQ. We have recorded some arguments that circulated on this dispute. Our analysis suggests that arguments values and interpretations may depend on the context in which they are exchanged. Arguments do not bear the same weight in all spaces because different communities may resort to several notions of proof, have different expectations with respect to errors, levels of uncertainty and acceptable time frames to remove such uncertainties.

2   Public discussion: different format for science and media

As a consequence of the planetary emergency, the mediatic world discussed and spread scientific knowledge in real time since the beginning of the crisis. However, at the beginning of the crisis in Europe, which can be dated back to February 2020, scientific knowledge on the SARS-CoV-2 virus was not yet present, but was being constructed, which is a usual moment for science to witness the emergence of controversies. Unlike most of the scientific controversies, the Covid-19 crisis showed in real-time how messy the scientific construction process is. And, unlike most of the scientific controversies, research findings were shared and made public as soon as available. The result is that two very different worlds collided: the one of media, whose temporality is rather short, with news to be published on an everyday basis, and the one of science, which can take decades to get to an important result and where the consensus around a theory is not necessarily a synonym for its truth.

Controversies are very important in the scientific world in periods when there are unknowns: they are defined by disagreements on results or general laws, or on methods to demonstrate. What makes a controversy, rather than a mere disagreement, is the fact that there is a real long lasting discussion where different trends of research are opposed, and often disagree on the status of the proof that each side is proposing. Solving controversies means that one gets to a representation of the world that can be consensual, and can lead to useful applications.

3   Scientific logic

In a world of science that would be pure of any other type of influence, ideally, this convergence emerges after a long academic dispute. This dispute is based on the exchange of arguments, each one being defined by clear paradigms, methods, measures obtained thanks to these methods and interpretation, with the limits that help define the level of truth that can be applied to this overall acquired knowledge. "Refutability" is the word that is commonly used to refer to this research of transparency and the acceptation of open critics and discussion to revise a proof or analysis:

> Science and scientific objectivity do not (and cannot) result from the attempts of an individual scientist to be 'objective,' but from the friendly-hostile cooperation of many scientists. [3]

One important element that enables to engage in such a process, is the charity principle [4], which means that one should always assume that the other human is as rational as oneself and try to understand its rationality instead of projecting our own understanding.

When considering the Covid-19 time, after one year, there are still many open discussions, like whether an early cures exist and are efficient and what are the legitimate methods to prove the efficiency of a cure, how efficient is lockdown on the dynamics of epidemics and on mortality rates, etc. Interestingly, many people have the belief that these different questions are solved scientifically, whereas they have been solved socially – in administrative norms or in the mediatic world – but are still discussed within the scientific community.

4 Error for science vs media

Unlike scientists, journalists have to provide the public with pieces of information on an everyday or weekly basis, and belong in general to a media that has some political preferences, which impose a special framing to the interesting topics. But there are two instruments to found deontology of journalism: the information has to be linked to a source that can be checked again, be it data, a person, an institution; a serious article about a debate should cite sources from both opposing sides, so that the public knows about the type of arguments and beliefs.

Another aspect that differs from science is that a mistake is considered as a sin: if someone makes a mistake, then she is considered as unreliable for the following period. This idea does not make the difference between different ways of being wrong in science:

- When predicting something, it is very easy to be wrong in science, because prediction is usually based on models, that are defined thanks to reductionist assumptions. But the world is complex: most of the time a non-modelized element makes the model wrong in its predictions. This does not mean the model is always wrong, but maybe that it should not be used for prediction.

- When describing / explaining an event, again it is possible to be wrong because of some reductionist assumptions or due to some missing explanatory elements. The mistake is beneficial to the scientist because she can then revise the space of applicability of the model;

- Eventually the model can be proven wrong in so many applications that it could be good to revise it – but there is a certain lag between the recognition that a model is not useful and its disappearing. In this case, it is difficult to know if it is the scientist who is wrong or if it is the theory (or the model).

What is important for a scientist is thus not to avoid errors, but to be able to recognize the error and revise either her way of applying the model, or the theory itself. This humble attitude is necessary to produce trust from colleagues, which is the condition of the acceptance of the normality of error.

In media, the expectations regarding the scientist is that she knows more than usual people, but should also know without errors. Science is not seen as a journey towards the truth, but as a revelation of truth. This religious relation to science explains the tendency to excommunicate publicly certain scientists, or consider others as knowing the definitive truth.

5   An example during Covid Crisis: discovering HCQ effects

As mentioned in the Introduction, in this paper we focus on the dispute in France over the effects of the hydroxychloroquine (HCQ) treatment on the SARS-CoV-2 virus. We can reconstruct the debate over several steps:

- **What is an acceptable proof?** The first results announced by Didier Raoult were refuted because these results came from a very small set of individuals, so they were just an indication, not a proof. Raoult's response was that, if a treatment was working, it was more important to take care of the patients and let others proving the efficacy of the treatment. If Raoult was supported by some in the media, the opposing view was that a Randomized controlled trial (RCT) was necessary to prove the treatment: the medicine has to be proven better than placebo, better than any other used molecule. Raoult wrote then a tribune to support his method with the argument that the re-usability of an old molecule is a bet that should be done (and financed by States). He also attacked the RCT, on the basis that they are expensive experiments, thus requiring to be financed by the pharmaceutical industry, insinuating they are certainly biased. Later he added the argument that the number of people that have to be enrolled in RCT for a disease that kills 0,5% poses some serious ethical concerns [1]: 40000 patients have to be treated with the molecule and 40000 without. Interestingly, the media regarded RCT as the "gold standard". However, later some research [2] showed that one RCT cannot prove more than any other statistical study. It is a proof of non-rejection of an hypothesis and not the proof that the hypothesis is right.

- **Which side are you?** The public opinion on social networks, journalists and 'experts' on TV shows all take position for (using the 'result' argument) or against Raoult (using the 'method' argument). According to one poll conducted for LCI at the end of May 2020, 45% of French people trust Didier Raoult and 35% have a bad opinion of him. The discussion around him led people to remove friends from their Facebook.

- **Lancet or not Lancet?** A paper published in the prestigious medical journal the Lancet showed that the combined treatment with HCQ and AZ kills people. This resulted in the halting of trials of HCQ in the French trial Discovery. However, two weeks later, the Lancet retracted the paper, described by his editor as "a shocking example of research misconduct". The fraud was so obvious that even people who did not agree with Raoult defended him. Although the Lancet paper was retracted, journalists still think HCQ is dangerous (and, so, Raoult is a dangerous man). Although papers are still produced at high rate and can be accessed quite easily (https://c19hcq.com), in most media discourses people who still wonder if (and how) HCQ can work on SARS-CoV-2 infections are judged silly. Without knowing what future will tell, if so many papers get published by diverse researchers, it means that the dispute is not over. These two worlds produce acceptable argument sets that are disjoined.

References

[1] J.O. Ovosi, M. I. and B. Bello-Ovosi, *Randomized controlled trials: Ethical and scientific issues in the choice of placebo or active control*, Annals of African Medicine **16(3)** (2017), pp. 97–100.
[2] Mielke, D. and V. Rohde, *Randomized controlled trials – a critical re-appraisal*, Neurosurgical Review (2020).
[3] Popper, K., "The Open Society and its Enemies," London: Routledge, 1945.
[4] Quine, W., "Word and Object," Cambridge, Mass., MIT Press, 1960.

A Modal Logic of Defeasible Reasoning

Huimin Dong, Yì N. Wáng [1]

{*huimin.dong,ynw*}*@xixilogic.org*

Department of Philosophy (Zhuhai), Sun Yat-Sen University

Abstract

We propose a default modal logic for defeasible reasoning by modeling defaults using the notions of consistency and preference. A default sentence "from φ, presuming the consistency of χ, normally infer ψ" is interpreted by – in one of the versions – "ψ is true in all the least exceptional φ-worlds where χ is possible". We use an alethic modal operator for characterizing the consistency/possibility of presumptions, and a default inference is characterized by a default modal implication with the normality and exceptionality interpreted using a binary relation. We study the resulting logic and discuss possible generalizations and relationships to the literature.

Keywords: defeasible reasoning, default logic, modal logic, justification, preference.

1 Introduction

The studies of defeasible reasoning [11,17] investigate the ways of inferring information when additional evidences are given, such that the monotonic property in classical logic is not satisfied. Given a set of premises, in classical or monotonic reasoning, a conclusion is preserved even when more information is added to the premises. In defeasible reasoning, conclusions are uncertain: they are sensible to and may be defeated by the new pieces of information.

Many logical theories and methods have been proposed to study defeasible and nonmonotonic reasoning [17], including the early work in the 1980s, such as default logics [18,1], circumscription [12,13], autoepistemic logic [14], and later developments, for instance conditional logics on various preferences [5,8,4] and their dynamics [22,21], defeasible logics [15], adaptive logics [20], and input/output logics [10]. In this paper we focus on default logics, and develop a modal logic for defeasible reasoning based on default rules.

Default logics are rule-based systems to reason about uncertain information, adopting a set of the so-called *default rules* to capture uncertainty. For a given language, a default rule or simply a *default* is an inference rule r of the following type:

$$\frac{\varphi : \chi_1, \ldots, \chi_n}{\psi}$$

---

[1] Corresponding author.

where φ is a formula called the *premise* or *prerequisite* of r, formulas $\chi_1, \dots, \chi_n$ are called the *justifications*, and the formula ψ is the *consequent* of r.

Default rules well represent many sentences of uncertain information in natural language. Consider the following conditional:

If it is Sunday then I will go fishing, unless I wake up late or my parents visit me.

Given the truth of the antecedent, the conclusion will be true, if it is not in a certain exceptional case declared by the unless-clause. This sentence is usually represented in terms of the following default:

$$ r = \frac{Sunday \ : \ \neg wake\text{-}up\text{-}late, \ \neg parents\text{-}visit}{fishing} $$

where the sentence *Sunday* stands for "it is Sunday", *fishing* for "I will go fishing", *wake-up-late* for "I wake up late (on Sunday)", and *parents-visit* for "my parents visit me (on Sunday)." In this case, *Sunday* is seen as the premise of the default r, and *fishing* the consequent. When the negation of the exception *wake-up-late* and *parents-visit* are both consistent with the premise, the default rule r can be applied in the process of reasoning.

Looking closer into the interpretation of the default rule illustrated above, we find that while the *truth* of the premise and consequent is essential for the default reasoning, the truth of the justifications is not. It is the *consistency*, or *possibility* viewed in a modal perspective, of the justifications that matters here. In other words, our view of the default rule r is as follows:

$$ r = \frac{Sunday \ : \ \Diamond\neg wake\text{-}up\text{-}late, \ \Diamond\neg parents\text{-}visit}{fishing} $$

where $\Diamond\varphi$ represents that φ is consistent (with the premise) and will be treated as a standard modal diamond operator interpreted in Kripke semantics [3].

We introduce a bimodal language with a standard $\Diamond$ operator serving the above purpose, together with a modal connective $\rightsquigarrow$ for *default implication* that characterizes the "if $\cdots$ then normally $\cdots$" clause in a defeasible way. In this language we can express different types of defaults, for example, the conditional above can then be written in our language, in one of the types, by

$$ Sunday \rightsquigarrow (fishing \wedge \Diamond\neg wake\text{-}up\text{-}late \wedge \Diamond\neg parents\text{-}visit). $$

The modal logic based on this language gives us a different but closely relevant approach to default reasoning.

The paper is structured as follows. In Section 2 we introduce a default modal logic based on the bimodal language for defeasible reasoning. On top of the logic we study different types of defaults in Section 3. We discuss possible generalizations and related work in Sections 4 and 5, respectively. We conclude in Section 6.

2 The logic DML

In this section we introduce a default modal logic (DML). We assume a set *Prop* of propositional variables. By representing default rules, we introduce two modal operators, one is a unary modality $\Diamond$ and the other a binary modality $\rightsquigarrow$. The former is used to capture consistency of justifications, while the latter is used for default implication. A default rule will be expressed by a compound formula using these operators (see Section 3).

Definition 2.1 The language $\mathcal{L}$ is defined as follows:

$$\varphi := p \mid \neg\varphi \mid (\varphi \to \varphi) \mid (\varphi \rightsquigarrow \varphi) \mid \Diamond\varphi$$

where $p \in Prop$. Other propositional connectives, such as $\vee$, $\wedge$ and $\leftrightarrow$, are defined in a standard way. $\Box\varphi$ is a shorthand for $\neg\Diamond\neg\varphi$.

A formula $\Diamond\varphi$ reads "φ is possible", which will be used to express the consistency or non-exception of φ. A formula $\varphi \rightsquigarrow \psi$ is called a *default implication*, which is introduced for representing a default conditional "from φ normally infer ψ". Now we define the formal models for DML.

Definition 2.2 A *model* is a tuple $M = (W, S, \preceq, V)$, where

- W is a non-empty set of (possible) worlds;
- S is a serial relation on W;
- $\preceq$ is a binary relation on W;
- $V : Prop \to 2^W$ is a valuation.

A *frame* is a model minus the valuation function.

While the general conventions of Kripke models for modal logic apply here, the relations S and $\preceq$ are introduced for modeling the consistency of justifications and properties of default implications.

Given worlds w and u, by wSu we mean that u is an evidence for justifying the formulas true in w. In such a case, $\Diamond\varphi$ is true in w, meaning that φ is consistent with formulas considered true in w, and formally this is the case if and only if there is a world u in which φ is true.

On the other hand, $w \preceq u$ indicates that all formulas in w are less exceptional than those in u. Note that we do not assume $\preceq$ to be reflexive or transitive here (see a discussion in Section 5).

We define $w \prec u$ to be $w \preceq u$ and $u \not\preceq w$, and define $w \sim u$ to be $w \preceq u$ and $u \preceq w$. Given a set X of worlds, a *minimal set* of X, denoted $\min_\prec(X)$, is the set $\{w \in X \mid \text{there is no } u \in X \text{ such that } u \prec w\}$. Each $w \in \min_\prec(X)$ can be understood as a set of formulas that are least exceptional on X.

The truth conditions are defined formally as follows.

Definition 2.3 Given a model $M = (W, S, \preceq, V)$, the satisfaction/truth of a formula in a world $w \in W$ is defined inductively below:

$$
\begin{aligned}
M, w &\models p & &\Longleftrightarrow & &w \in V(p) \\
M, w &\models \neg\varphi & &\Longleftrightarrow & &\text{not } M, w \models \varphi \\
M, w &\models (\varphi \to \psi) & &\Longleftrightarrow & &M, w \models \varphi \text{ materially implies } M, w \models \psi \\
M, w &\models (\varphi \rightsquigarrow \psi) & &\Longleftrightarrow & &\min_{\preceq}(\|\varphi\|) \subseteq \|\psi\| \\
M, w &\models \Diamond\varphi & &\Longleftrightarrow & &\text{there exists } u \in W \text{ such that } wSu \text{ and } M, u \models \varphi.
\end{aligned}
$$

where $\|\varphi\|$ is the truth set of φ (in M, w), i.e., $\{w \in W \mid M, w \models \varphi\}$.

Lemma 2.4 *Given a model* $M = (W, S, \preceq, V)$, *if* $Y \subseteq X \subseteq W$ *and* $w \in \min_{\preceq}(X) \cap Y$, *then* $w \in \min_{\preceq}(Y)$.

Proof. Let $w \in \min_{\preceq}(X) \cap Y$. For any $u \in X$, we know $u \not\prec w$. By $Y \subseteq X$, if $u \in Y$ then $u \not\prec w$. Since $w \in Y$, $w \in \min_{\preceq}(Y)$. $\qquad\square$

Lemma 2.5 *The following formulas are valid:*

- *Abs.* $(\varphi \rightsquigarrow \psi) \leftrightarrow \Box(\varphi \rightsquigarrow \psi)$
- *Sh.* $((\varphi \wedge \psi) \rightsquigarrow \chi) \to (\varphi \rightsquigarrow (\psi \to \chi))$

Proof. (Abs) is valid, due to the fact that the set $\min_{\preceq}(\|\varphi\|)$ does not vary among states of a given model.

As for the validity of (Sh), let $\min_{\preceq}(\|\varphi \wedge \psi\|) \subseteq \|\chi\|$, and suppose $M, w \not\models \varphi \rightsquigarrow (\psi \to \chi)$. Then there exists $u \in \min_{\preceq}(\|\varphi\|)$ such that $M, u \not\models \psi \to \chi$. So $u \in \|\psi\|$ and $u \notin \|\chi\|$ (@). From $u \in \min_{\preceq}(\|\varphi\|)$ and (@) we know that $u \in \|\varphi \wedge \psi\|$. By Lemma 2.4 we have $u \in \min_{\preceq}(\|\varphi \wedge \psi\|)$. It follows that $u \in \|\chi\|$, which contradicts (@). $\qquad\square$

The validity (Abs) reflects that minimality is a global/model-level property, which follows from the interpretation of the default implication (the case for $\varphi \rightsquigarrow \psi$ in Definition 2.3). The validity (Sh) is often seen in nonmonotonic logic. It is named after Shoham [19], and corresponds to the so-called *conditionalization* principle. This principle is part of the system P for the preferential consequence relation [9].

Proposition 2.6 *The following formulas and rules are valid in DML:*

- *Right weakening (RW): from* $\varphi \to \psi$ *inferring* $(\chi \rightsquigarrow \varphi) \to (\chi \rightsquigarrow \psi)$;
- *Cautious monotonicity (CM):* $(\varphi \rightsquigarrow \psi) \wedge (\varphi \rightsquigarrow \chi) \to ((\varphi \wedge \psi) \rightsquigarrow \chi)$;
- *OR:* $(\varphi \rightsquigarrow \chi) \wedge (\psi \rightsquigarrow \chi) \to ((\varphi \vee \psi) \rightsquigarrow \chi)$;
- *AND:* $(\varphi \rightsquigarrow \psi) \wedge (\varphi \rightsquigarrow \chi) \to (\varphi \rightsquigarrow (\psi \wedge \chi))$;
- *Constrained modus ponens (CMP):* $(\varphi \rightsquigarrow \psi) \wedge (\varphi \rightsquigarrow (\psi \to \chi)) \to (\varphi \rightsquigarrow \chi)$.

Proposition 2.7 *The following formulas or rules are not valid in DML:*

- *EHD:* $((\varphi \rightsquigarrow (\psi \to \chi)) \to ((\varphi \wedge \psi) \rightsquigarrow \chi))$;
- *Transitivity:* $(\varphi \rightsquigarrow \psi) \wedge (\psi \rightsquigarrow \chi) \to (\varphi \rightsquigarrow \chi)$;
- *Contraposition:* $(\varphi \rightsquigarrow \psi) \to ((\neg\psi) \rightsquigarrow (\neg\varphi))$;
- *Monotonicity: from* $\varphi \to \psi$ *inferring* $(\psi \rightsquigarrow \chi) \to (\varphi \rightsquigarrow \chi)$.

The axiomatization **DML** is given as follows.

(PC)	instances of propositional tautologies
(dual)	$\Box\varphi \leftrightarrow \neg\Diamond\neg\varphi$
(K)	$\Box(\varphi \rightarrow \psi) \rightarrow (\Box\varphi \rightarrow \Box\psi)$
(D)	$\Box\varphi \rightarrow \Diamond\varphi$
(Dist)	$(\chi \rightsquigarrow (\varphi \rightarrow \psi)) \rightarrow ((\chi \rightsquigarrow \varphi) \rightarrow (\chi \rightsquigarrow \psi))$
(L-Ext)	from $(\varphi \leftrightarrow \psi)$ infer $((\varphi \rightsquigarrow \chi) \rightarrow (\psi \rightsquigarrow \chi))$
(R-Ext)	from $(\varphi \leftrightarrow \psi)$ infer $((\chi \rightsquigarrow \varphi) \rightarrow (\chi \rightsquigarrow \psi))$
(Id)	$\varphi \rightsquigarrow \varphi$
(Abs)	$(\varphi \rightsquigarrow \psi) \leftrightarrow \Box(\varphi \rightsquigarrow \psi)$
(Sh)	$((\varphi \wedge \psi) \rightsquigarrow \chi) \rightarrow (\varphi \rightsquigarrow (\psi \rightarrow \chi))$
(MP)	from φ and $(\varphi \rightarrow \psi)$ infer ψ
(Gen)	from φ infer $\Box\varphi$

Theorem 2.8 *The proof system* **DML** *is sound for DML.*

Two sound and strongly complete systems are given for conditionals $\varphi \rightsquigarrow \psi$ which are interpreted by reflexive, respectively, reflexive and transitive relations within the assumption of smoothness [16].

3   Defaults in DML

In this section we study default rules in the framework of DML.

3.1   Applicability of defaults

Given a default $r = \dfrac{\varphi : \chi_1, \ldots, \chi_n}{\psi}$, we shall write (i) $pre(r)$ for the premise of r, i.e., $pre(r) = \varphi$, (ii) $cons(r)$ for the consequent of r, i.e., $cons(r) = \psi$, and (iii) $just(r)$ for the justifications of r, i.e., $just(r) = \{\chi_1, \ldots, \chi_n\}$.

The closure $cl(R)$ of a set R of default rules is the set of formulas consisting of all premises and consequents of rules in R, i.e., $cl(R) = \{pre(r), cons(r) \mid r \in R\}$. The part of R that is outside $cl(R)$ is denoted $just(R)$. Namely, $just(R) = \bigcup_{r \in R} just(r)$ is the set of all justifications of rules in R. Put in a different way, the pair $\langle cl(R), just(R) \rangle$ partitions R into two main components.

Applicability of defaults is one of the key concepts driving the development of various default logics [1]. Roughly speaking, a set R of defaults is *applicable* if certain constraints on $\langle cl(R), just(R) \rangle$ are satisfied. Different sets of constraints result in different default logics. Here we give a simple example to illustrate the idea, and we refer to [1] for more details.

Example 3.1 Consider the following three defaults from [1, Section 7.4]:

$$ r_1 = \frac{\top : p}{q} \qquad r_2 = \frac{\top : \neg p}{r} \qquad r_3 = \frac{\top : \neg q, \neg r}{s} $$

There are various ways to consider whether certain subsets of $\{r_1, r_2, r_3\}$ are applicable, depending on different principles of consistency defined in terms of the justifications of the defaults. For instance, under the principle requiring that the justifications of every single default does not contradict with the closure of $\{r_1, r_2, r_3\}$ (i.e., $cl(\{r_1, r_2, r_3\})$, namely, all the premises and consequents of r_1, r_2 and r_3), the set $\{r_1, r_2\}$ is applicable (the justifications of r_3 contradicts with the consequent of r_1). When a more rigorous principle, requiring that the justifications of these defaults altogether be consistent, the singletons $\{r_1\}$, $\{r_2\}$ and $\{r_3\}$ are applicable but none of their unions.

In different default logics, different principles are employed to identify various defaults as being capable to be applied together [1]. Below we present three accounts briefly. Given a set R of defaults.

Principle I We consider a set R' of defaults applicable when every element of $just(R')$ is consistent with $cl(R)$ and there is no $R'' \supset R'$ such that every element of $just(R'')$ is consistent with $cl(R)$. In the example, the set $\{r_1, r_2\}$ is typically recognized as the unique set of applicable rules.

Principle II We consider a set R' of defaults applicable when every element of $just(R')$ is consistent with $cl(R')$ and there is no $R'' \supset R'$ such that every element of $just(R'')$ is consistent with $cl(R'')$. The sets $\{r_1, r_2\}$ and $\{r_3\}$ in the example are both considered as being applicable.

Principle III We consider a set R' of defaults applicable when $just(R') \cup cl(R')$ is consistent and there is no $R'' \supset R'$ such that $just(R'') \cup cl(R'')$. The singletons $\{r_1\}$, $\{r_2\}$ and $\{r_3\}$ in the above example are treated as sets of applicable rules in this case.

Each principle attempts to minimize the exceptional cases of default application. These three accounts provide ways to keep the desired sets of defaults *as less exceptional as possible*.

Clearly, the partition $\langle cl(R), just(R) \rangle$ on R offers a ground to investigate various default logics. The above observation brings to light that *minimality* and *consistency* based on this partition are two key components to decide applicable default rules. We model this idea in the next section.

In terms of DML, we say a default implication $\varphi \rightsquigarrow \psi$ (without an exception declared explicitly) is applicable at w when this formula is satisfied at w. Simply to say, *applicability is identified as satisfaction* in DML.

Example 3.2 We model Example 3.1 as follows. Let $M = (W, S, \preceq, V)$ be a model such that:

- $W = \{w_1, w_2, w_3, w_4\}$,
- $S = \{(w_1, w_3), (w_1, w_4), (w_2, w_2), (w_3, w_3), (w_4, w_4)\}$,
- $w_1 \prec w_3 \sim w_4 \prec w_2$, and
- $V(p) = \{w_3\}$, $V(q) = \{w_1, w_2\}$, $V(r) = \{w_1, w_4\}$, and $V(s) = \{w_2\}$.

Let $R = \{r_1, r_2, r_3\}$.

The model M from Example 3.2 (pictured partially in Figure 1, with the order $\preceq$ made implicit) illustrates three Principles for applicability of default rules in Example 3.1.

First, each world illustrates a closure of a particular subset of R. The conjunction $(q \wedge r)$ of the closure of $\{r_1, r_2\}$ is true at w_1, $(\top \wedge s)$ of the closure

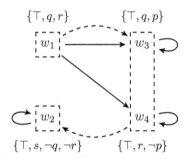

$\{\top, q, r\}$ $\{\top, q, p\}$

$\{\top, s, \neg q, \neg r\}$ $\{\top, r, \neg p\}$

Fig. 1. Illustration of the model M in Example 3.2. Dashed squares indicate worlds that are equally exceptional. A dashed arrow leads from less exceptional worlds to more exceptional ones. Solid arrows indicate the relation S.

of $\{r_3\}$ is true at w_2, $(\top \wedge q)$ of the closure of $\{r_1\}$ is true at w_3, and the conjunction $(\top \wedge r)$ of the closure of $\{r_2\}$ is true at w_4. Then, we simply consider each world as the set of rules whose closures are true there. This fits well to our intuition of instantiating defaults at possible worlds.

Second, all the justifications are captured by the relation S. The formula $(\Diamond p \wedge \Diamond \neg p)$ represents that p and $\neg p$ are the justifications of $\{r_1, r_2\}$, and this conjunction is true at w_1. While, $\Diamond p$ as the unique justification of $\{r_1\}$ is true at w_3 and $\Diamond \neg p$ as that of $\{r_2\}$ is true at w_4. Further, $(\Diamond \neg q \wedge \Diamond \neg r)$ as the justifications for $\{r_3\}$ is true at w_2. Since the relation S is required to be serial, the consistency between justifications can be displayed naturally.

The relation $\preceq$ adopts a quantitative way of evaluating justifications: a world is less exceptional than another if the former has more justifications that are consistent with the closure of R than the latter. w_1 has two justifications, as $\Diamond p$ and $\Diamond \neg p$ are both true at w_1. Only $\Diamond p$ or $\Diamond \neg p$ is true at w_3 and w_4. They are strictly more exceptional than that of w_1, and at the same time equally exceptional to each other. Both $\Diamond \neg q$ and $\Diamond \neg r$ true at w_4 are not consistent with the closure of R, and then w_2 is strictly more exceptional than all others.

A *default implication* in terms of DML provides us a ground to express default implication in the sense of whether the premise *normally* implies the consequent. For example, the formulas $(\top \rightsquigarrow q)$ and $(\top \rightsquigarrow r)$ are true at every world of M in the above example, and on the other hand, no world of M satisfies $(\top \rightsquigarrow s)$. Regarding the applicability of defaults with justifications being considered, the three principles for applicability can be also expressed in DML. We shall study this in the following section.

3.2 Variants of defaults

In DML we have introduced the modal operator for default implication, $\rightsquigarrow$, to describe conditionals such as "If it is Sunday then I normally will go fishing." Justifications in a default rule can be expressed by a conjunction of $\Diamond$-sentences. As discussed in the previous section, various default logics are developed according to different understandings of the applicability of default rules. In this section, we define variants of defaults in DML, to *explicitly* express their logical mechanisms underlying the notion of applicability.

Dominant default Given a model $M = (W, S, \preceq, V)$ and a world $w \in W$, "If φ then ψ, with the justifications $\chi_1, \ldots,$ and χ_n" is a *dominant default*, denoted $(\varphi \xrightarrow{\chi_1, \ldots, \chi_n}_d \psi)$, if the following holds:

$$M, w \models (\varphi \xrightarrow{\chi_1, \ldots, \chi_n}_d \psi) \iff \min_{\preceq}(\|\varphi\|) \subseteq \|\psi\| \cap \bigcap_{1 \leq i \leq n} \|\Diamond \chi_i\|.$$

It is not hard to observe that the dominant default $\varphi \xrightarrow{\chi_1, \ldots, \chi_n}_d \psi$ can be expressed in DML by the formula $(\varphi \rightsquigarrow (\psi \wedge \bigwedge_{1 \leq i \leq n} \Diamond \chi_i))$.

A dominant default indicates that, when the premise of such a default is true at a least exceptional world, its justifications must all be consistent with such a minimal state.

The relation $\preceq$ in Example 3.2 indicates the consistency between the instantiated justifications and the closure of all rules. Then, a dominant default valid in this model is the one whose justifications must be consistent with all closures of R as much as possible. The formulas $(\top \xrightarrow{p}_d q)$ and $(\top \xrightarrow{\neg p}_d r)$ are true in all states of the model M from Example 3.2.

Global default Given a model $M = (W, S, \preceq, V)$ and a world $w \in W$, "If φ then ψ, with the justifications $\chi_1, \ldots,$ and χ_n" is a *global default*, denoted $(\varphi \xrightarrow{\chi_1, \ldots, \chi_n}_g \psi)$, if the following holds:

$$M, w \models (\varphi \xrightarrow{\chi_1, \ldots, \chi_n}_g \psi) \iff \min_{\preceq}(\|\varphi\| \cap \bigcap_{1 \leq i \leq n} \|\Diamond \chi_i\|) \subseteq \|\psi\|$$

Observe that the global default $(\varphi \xrightarrow{\chi_1, \ldots, \chi_n}_g \psi)$ can be expressed by the formula $((\varphi \wedge \bigwedge_{1 \leq i \leq n} \Diamond \chi_i) \rightsquigarrow \psi)$ in DML.

A global default emphasizes that, when the premise and justifications of a given default are least exceptional at a state, its consequent can be concluded.

The global defaults $(\top \xrightarrow{p}_g q)$, $(\top \xrightarrow{\neg p}_g r)$ and $(\top \xrightarrow{\neg q, \neg r}_g s)$ are true in all the worlds of the model M from Example 3.2.

This notion of global default already provides a different perspective of valid defaults from that of dominant default. We consider a fine-grained version to identify applicable defaults locally.

Local default Given a model $M = (W, S, \preceq, V)$ and a world $w \in W$, "If φ then ψ, with the justifications $\chi_1, \ldots,$ and χ_n" is a *local default*, denoted $(\varphi \xrightarrow{\chi_1, \ldots, \chi_n}_l \psi)$, if the following holds:

$$M, w \models (\varphi \xrightarrow{\chi_1, \ldots, \chi_n}_l \psi) \iff w \in \| \bigwedge_{1 \leq i \leq n} \Diamond \chi_i \| \& M, w \models (\varphi \xrightarrow{\chi_1, \ldots, \chi_n}_g \psi)$$

Now we have $(\varphi \xrightarrow{\chi_1, \ldots, \chi_n}_l \psi)$ expressible by $((\varphi \xrightarrow{\chi_1, \ldots, \chi_n}_g \psi) \wedge \bigwedge_{1 \leq i \leq n} \Diamond \chi_i)$.

Local default is a notion relative to a possible world. Applying a local default not only needs to ensure that the default is *globally* applicable when its justification is least exceptional, but it also requires that the justification holds *locally* at the current world.

The following local defaults are true in the given world of the model M from Example 3.2:

- w_1: $(\top \xrightarrow{p}_l q)$, $(\top \xrightarrow{\neg p}_l r)$, $\neg(\top \xrightarrow{\neg q, \neg r}_l s)$;
- w_3: $(\top \xrightarrow{p}_l q)$, $\neg(\top \xrightarrow{\neg p}_l r)$, $\neg(\top \xrightarrow{\neg q, \neg r}_l s)$;
- w_4: $\neg(\top \xrightarrow{p}_l q)$, $(\top \xrightarrow{\neg p}_l r)$, $\neg(\top \xrightarrow{\neg q, \neg r}_l s)$;
- w_2: $\neg(\top \xrightarrow{p}_l q)$, $\neg(\top \xrightarrow{\neg p}_l r)$, $(\top \xrightarrow{\neg q, \neg r}_l s)$.

Self-maintained default Given a model $M = (W, S, \preceq, V)$ and a world $w \in W$, "If φ then ψ, with the justifications χ_1, ..., and χ_n" is a *self-maintained default*, denoted $(\varphi \xrightarrow{\chi_1, \ldots, \chi_n}_s \psi)$. Its truth condition is given by, $M, w \models (\varphi \xrightarrow{\chi_1, \ldots, \chi_n}_s \psi)$ if and only if

(i) there exists $u \in W$ such that $M, u \models \bigwedge_{1 \leq i \leq n} \Diamond \chi_i$, and

(ii) $\min_{\preceq}(\|\varphi\|_{\chi_1, \ldots, \chi_n}) \subseteq \|\psi\|_{\chi_1, \ldots, \chi_n}$,

where $\|\varphi\|_{\chi_1, \ldots, \chi_n} = \{u \in M_{\chi_1, \ldots, \chi_n} \mid M_{\chi_1, \ldots, \chi_n}, u \models \varphi\}$ and $\|\psi\|_{\chi_1, \ldots, \chi_n}$ defined likewise, in which the updated model $M_{\chi_1, \ldots, \chi_n} = (W^*, S^*, \preceq^*, V^*)$ is such that:

- $W^* = \{u \in W \mid M, u \models \chi_1 \wedge \cdots \wedge \chi_n\}$;
- S^* is $S \cap (W^* \times W^*)$ with all its dead ends equipped with a self-loop (to make sure that S^* is serial);
- $\preceq^* = \preceq \cap (W^* \times W^*)$;
- $V^*(p) = V(p) \cap W^*$, for all atoms p.

The consistency of closures is already warranted by the notion of possible worlds. By ensuring the consistency of justifications everywhere in an updated manner, self-maintained default makes sure all applicable rules having consistency among their closures together with their justifications.

The formulas $(\top \xrightarrow{p}_s q)$, $(\top \xrightarrow{\neg p}_s r)$ and $(\top \xrightarrow{\neg q, \neg r}_s s)$ for self-maintained defaults are true in all worlds of the model M from Example 3.2.

Proposition 3.3 *The above four types of default are different.*

Back to the previous discussion on the Principles I, II and III in Section 3.1, they can be *explicitly* expressed in terms of the language of DML. The three principles of defining applicable defaults can be captured respectively by dominant, local and self-maintained defaults. Briefly speaking, the three concepts of defaults coincide with those three different principles of handling consistency between justifications and closures. To make this coincidence clear it would require some space, and we have to leave this out here.

4 Generalization

The default implication, $\varphi \rightsquigarrow \psi$, that we have considered in the paper is interpreted irrespectively of the factual world, leading to the axiom (Abs). This is intended to characterize an *objective* version of default reasoning. We can also introduce a variant for *subjective* defaults, so that defaults can vary among possible worlds. This is easy to do. In a model we can introduce a binary relation $\preceq$ relative to possible worlds. Formally, a *subjective model* is a tuple $M = (W, R, \preceq, V)$, such that W, R and V are the same as in a (objective) model (Definition 2.2) and $\preceq: W \to 2^{W \times W}$ assigns to every possible world w a binary relation on W (which is denoted $\preceq_w$). A subjective default implication $\varphi \rightsquigarrow^s \psi$ can then be interpreted as follows:

$$M, w \models (\varphi \rightsquigarrow^s \psi) \quad \Longleftrightarrow \quad \min_{\preceq_w}(\|\varphi\|) \subseteq \|\psi\|.$$

There are more general frameworks in the literature along the tradition of using conditionals to capture defaults. A version of default modal logic was proposed in [2], where a default rule is treated as a special type of modal connectives, denoted $\xrightarrow{\Box}$, in a modal language. A default then is characterized by an implication $\varphi \xrightarrow{\Box} \psi$, which reads "$\varphi$ normally implies ψ". Such an implication is interpreted using a *filter-based model* $F = (W, N, V)$ with:

- W is a non-empty set of (possible) worlds,
- $N : W \to 2^W \to 2^{2^W}$ assigning to every world $w \in W$ and to every set of worlds $A \subseteq W$, a family of sets of worlds, denoted $N_w(A)$, such that $A \in N_w(A)$ and $N_w(A)$ is a filter on W, [2]
- $V : Prop \to 2^W$ is a valuation,

such that for any world $w \in W$, $F, w \models \varphi \xrightarrow{\Box} \psi$ iff $\|\psi\| \in N_w(\|\varphi\|)$.

We can view the subjective default implication $\varphi \rightsquigarrow^s \psi$ as a special case of $\varphi \xrightarrow{\Box} \psi$, in the following sense.

(i) On one hand, given a subjective model $M = (W, R, \preceq, V)$, we can come up with a filter-based model $F^M = (W, N, V)$ such that for any $w \in W$ and $A \subseteq W$, $N_w(A)$ is the *principal filter* generated by $\min_{\preceq_w}(A)$. [3]

(ii) On the other hand, given a filter-based model $F = (W, N, V)$ such that for every $w \in W$ and $A \subseteq W$, $N_w(A)$ is a principal filter on W generated by $A \subseteq W$. We can define a subjective model $M^F = (W, R, \preceq, V)$ such that R is arbitrary and for every $w \in W$ and $A \subseteq W$, $\min_{\preceq_w}(A)$ is the smallest set in $N_w(A)$ (i.e., the unique, least filter base of $N_w(A)$).

We have the following result.

Theorem 4.1 *Let φ and ψ be arbitrary formulas.*

- *Given a subjective model M and a possible world w of M, $M, w \models \varphi \rightsquigarrow^s \psi$ iff $F^M, w \models \varphi \xrightarrow{\Box} \psi$;*
- *Given a filter-based model F and a possible world w of F, $F, w \models \varphi \xrightarrow{\Box} \psi$ iff $M^F, w \models \varphi \rightsquigarrow^s \psi$.*

The above theorem shows that the subjective models we have proposed above corresponds to the principal-filter-based models of [2], hence we get a

---

[2] A filter on a set W is a non-empty family $F \subseteq 2^W$ that is upward closed (if $A \in F$ and $A \subseteq B$ then $B \in F$) and closed under finite intersection (if $A, B \in F$, then $A \cap B \in F$).

[3] The principal filter (on W) generated by A (with $A \subseteq W$) is the family of all subsets of W containing A, i.e., $\{B \subseteq W \mid A \subseteq B\}$.

stronger logic than the latter. Furthermore, the logics we have proposed includes a type of modalities intended to characterize the justification of default, which was not studied in [2].

5 Related Work

Specificity for Preference Specificity can be used as a principle to develop preference-based models for defeasible inferences. Delgrande [6]'s preference model is one of these. Given a set of applicable rules, a preference is induced to determine which rules are best outcomes based on a set of contingent information. The way to define such a preference follows the *principle of specificity*: the possible worlds satisfying more applicable rules are less exceptional. An induced preference must be reflexive and transitive, but not necessarily comparable. In this case, a defeasible inference $\varphi \mathrel{\vdash_\Delta} \psi$ is defined as $\min_{\preceq_R}(\|\varphi\| \cap \|\Delta\|) \subseteq \|\psi\|$, where Δ is the background information and φ is the contingent and uncertain information. If our binary relation $\preceq$ in defined as a Delgrande's order, namely $\preceq := \preceq_R$, then $\Delta \xrightarrow{\varphi}_g \psi$ is equal to $\Diamond\varphi \mathrel{\vdash_\Delta} \psi$. In that case, Delgrande's solutions to the paradoxes [6] can be applied in ours. The important difference between Delgrande's work and ours is that our DML provides a general way to define the relation $\preceq$ to compare possible worlds. In Example 3.2, the relation $\preceq$, instead of satisfiying the principle of specificity, it links the worlds which contain more consistent justifications as less exceptional, and this relation gives us proper characterizations of three kinds of applicable defaults. If $\preceq$ is replaced by $\preceq_R$ of specificity, then $w_1 \prec_R w_3 \sim_R w_4$ but w_2 is not comparable with these three worlds. In such a model, $\top \xrightarrow{p}_d q$ and $\top \xrightarrow{\neg p}_d r$ are not true in all worlds and then cannot represent applicable defaults by Principle I. There are other principles of defining preferences (e.g. [5,4]). We leave the comparisons with these conditional logics of preferences for future work.

Variants of minimality A classical assumption for a mathematical structure of preference is that the set of minimal elements is non-empty. The assumption can be rephrased as: preferences are transitive. This can be lifted in terms of default rules, and, after that, variants of minimality can be considered. Consider the following three defaults:

$$r_4 = \frac{\top : p}{q} \qquad r_5 = \frac{\top : \neg q}{r} \qquad r_6 = \frac{\top : \neg r}{\neg p}$$

They can be instantiated by three possible worlds depicted in Figure 2, namely w_i instantiates only r_i ($i \in \{4, 5, 6\}$). The preference $\preceq$ is set in this way: a world w is less exceptional than u if and only if all justifications of the instantiated rules at w are consistent with the closures of the instantiated rules at u. In other words, the three worlds are ordered in a circle (see the dashed arrows): $w_4 \prec w_5 \prec w_6 \prec w_4$. In such a case, $\min_\preceq \|\top\|$ is empty, or $\preceq$ is non-transitive. To capture the *applicability* of defaults in this case, a new type of default implication needs to be proposed. Grossi et al. [7] examine four possible ways to model preferences by lifting this classical assumption In the

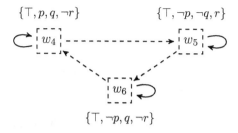

$$\{\top, p, q, \neg r\} \qquad\qquad \{\top, \neg p, \neg q, r\}$$

$$w_4 \qquad\qquad w_5$$

$$w_6$$

$$\{\top, \neg p, q, \neg r\}$$

Fig. 2. Illustration of a model.

future work we plan to adopt this idea in modeling applicable default rules in the above case.

Defeasible inferences based on system P A defeasible inference based on preferential structure is first introduced by [9]. This is often referred to as the KLM method, which is used to handle the reasoning from uncertain information to *plausible* conclusions. In the preferential KLM models, smoothness [9] [4] is required for the preferential relation $\preceq$ on W. A defeasible inference $\varphi \hspace{1pt}\vdash\hspace{-6pt}\sim \psi$ is true in a KLM-model when all the minimal worlds which make φ true also make ψ true. It is straightforward to consider the default implication $\rightsquigarrow$ in our paper as a representation of such a defeasible inference. The system P in KLM [9] is obtained by adding to the counterparts of DML axioms and rules (Id), (L-Ext), (RW), (CM) and (OR), together with the following rule, (Cut), the counterpart of which, however, is not valid in DML.

- Cut: from $\varphi \wedge \psi \hspace{1pt}\vdash\hspace{-6pt}\sim \chi$ and $\varphi \hspace{1pt}\vdash\hspace{-6pt}\sim \psi$ inferring $\varphi \hspace{1pt}\vdash\hspace{-6pt}\sim \chi$.

There are two derived principles that are also derivable in DML, namely (AND) and (CMP). The invalid formulas in Proposition 2.7 have counterexamples in the system P and many other systems for nonmonotonic reasoning.

6 Conclusion

We introduced a default modal logic (DML) that provides a general framework for characterizing various types of default reasoning, in particular, based on different criteria of applicability of default rules. We listed the corresponding expressions of these criteria, though without giving a proof for the correspondence. The framework is flexible enough to cover other types of defaults in literature, and we plan to look into this in the future.

We introduced a sound axiomatization for DML, with an intention to show that it is, or extend it to be, a complete one. This is, however, left for future work. We are also interested in the tableaux system and decision procedures for the logic.

The interpretation of the default implication was based on an *objective* version, and we generalized it to the *subjective* version which is seen as a stronger

---

[4] An alternative assumption is the well-foundedness, namely $\min_{\preceq}(X) \neq \emptyset$ for every $X \subseteq W$ and $X \neq \emptyset$.

variant of the logic proposed in [2]. It makes sense to study the subjective version in the future.

We had a brief comparison of DML with the system P, and argued that they are different. A further question would be whether the fragment of DML with only the modal connective for default implication (i.e, $\rightsquigarrow$, and without the modal $\Diamond$) is a instantiation of the system P.

Acknowledgments We would like to thank the three referees of LNGAI 2021 for their valuable remarks and comments. We acknowledge funding support by the National Social Science Fund of China (20CZX051, 20&ZD047).

References

[1] Antoniou, G., "Nonmonotonic Reasoning," The MIT Press, 1997.

[2] Ben-David, S. and R. Ben-Eliyahu-Zohary, *A modal logic for subjective default reasoning*, Artificial Intelligence **116** (2000), pp. 217–236.

[3] Blackburn, P., M. de Rijke and Y. Venema, "Modal Logic," Cambridge Tracts in Theoretical Computer Science **53**, Cambridge University Press, 2001, 554 pp.

[4] Boutilier, C., *Conditional logics of normality: a modal approach*, Artificial intelligence **68** (1994), pp. 87–154.

[5] Burgess, J. P. et al., *Quick completeness proofs for some logics of conditionals.*, Notre Dame Journal of Formal Logic **22** (1981), pp. 76–84.

[6] Delgrande, J., *A preference-based approach to defeasible deontic inference*, in: *17th International Conference on Principles of Knowledge Representation and Reasoning (KR 2020)* (2020), pp. 326–335.

[7] Grossi, D., W. Van Der Hoek and L. Kuijer, *Logics of preference when there is no best*, in: *17th International Conference on Principles of Knowledge Representation and Reasoning (KR 2020)*, 2020, pp. 455–464.

[8] Hansson, S. O., *Preference-based deontic logic (PDL)*, Journal of Philosophical Logic **19** (1990), pp. 75–93.

[9] Kraus, S., D. Lehmann and M. Magidor, *Nonmonotonic reasoning, preferential models and cumulative logics*, Artificial intelligence **44** (1990), pp. 167–207.

[10] Makinson, D. and L. van der Torre, *Constraints for input/output logics*, Journal of Philosophical Logic **30** (2001), pp. 155–185.

[11] McCarthy, J., *Some philosophical problems from the standpoint of artificial intelligence*, Machine Intelligence **4** (1969), pp. 463–502.

[12] McCarthy, J., *Epistemological problems of artificial intelligence*, in: *Proceedings of the 5th international joint conference on Artificial intelligence-Volume 2*, 1977, pp. 1038–1044.

[13] McCarthy, J., *Circumscription—a form of non-monotonic reasoning*, Artificial intelligence **13** (1980), pp. 27–39.

[14] Moore, R. C., *Semantical considerations on nonmonotonic logic*, Artificial intelligence **25** (1985), pp. 75–94.

[15] Nute, D., "Defeasible deontic logic," Dordrecht: Kluwer, 1997.

[16] Parent, X., *Maximality vs. optimality in dyadic deontic logic*, Journal of Philosophical Logic **43** (2014), pp. 1101–1128.

[17] Pollock, J. L., *Defeasible reasoning*, Cognitive science **11** (1987), pp. 481–518.

[18] Reiter, R., *A logic for default reasoning*, Artificial intelligence **13** (1980), pp. 81–132.

[19] Shoham, Y., "Reasoning about change," MIT press Cambridge, 1988.

[20] Straßer, C., "Adaptive logics for defeasible reasoning," Springer, 2014.

[21] van Benthem, J., D. Grossi and F. Liu, *Priority structures in deontic logic*, Theoria **80** (2014), pp. 116–152.

[22] Veltman, F., "Logics for conditionals," Ph.D. thesis, Universiteit van Amsterdam (1985).

Automated Translation of Contract Texts into Defeasible Deontic Logic

Luca Pasetto, Matteo Cristani

Department of Computer Science, University of Verona, Italy

Francesco Olivieri

Griffith University, Institute for Integrated and Intelligent Systems, Australia

Guido Governatori

Data61, CSIRO, Australia

Abstract

Many methods have been explored for translating legal texts into formal logic, but the results are yet far from being actually applicable to real-world problems, mainly because (a) natural language processing is intrinsically complex, (b) formalization of duties, prohibitions and permissions is a specific aspect of language processing that needs to be properly considered, and (c) legal texts often contain references to other texts.

We propose a methodology to analyse legal texts that represents an evolution of methods already devised in the literature and addresses the three aspects described above. We perform extraction of legal knowledge from a text containing an exploration permit taken from a corpus of resource contracts, and deploy it in the formal language of defeasible deontic logic.

Keywords: Legal knowledge, Defeasible Deontic Logic, Natural Language Processing

1 Introduction

Sources of legal knowledge are usually written in Natural Language, and in this way they form the **legal documents**. On the other hand, formalisms that can be treated by computational engines, such as *defeasible deontic logic*, lie at the opposite side of the line that goes from informal to formal structures that can be used to deploy juridically relevant information. Current methods to concretely transfer to formal language the knowledge contained in legal documents are of two kinds, both imperfect:

- Translation by hand, that can be rather accurate but definitely not sustainable in practice, as it requires a lasting effort by highly qualified personnel.

- Methods based on computer-based automation, that have been attempted in numerous studies, but that still give rather inaccurate, and in several cases erroneous, results.

Some investigations have proven that it is possible to devise a correct pipeline for the above mentioned goal, and that the accuracy limits of the automated translation can be overcome. In particular, starting from the pioneering work of Wyner and Peters [12] and subsequently that by Camilleri et al. [3], we can see a general line of improvement for these methods.

However, although these methods have shown some positive progress over the years, there are still several issues to solve in order to provide an appropriate overall method for the translation process. In particular, this research addresses the technical aspects of the translation from natural sentences in legal documents to formal language that emerged in the studies cited above, but remained open.

When we process a legal document to formalise it, we start by executing the basic operations of the typical pipeline of natural language processing, in summary: (1) Tokenisation, (2) Part-of-speech tagging, (3) Syntactic tree generation, (4) Translation into formal language.

One well-known source of computational complexity of the above process is Step (3), that is due, in turn, to the possible syntactic/semantic ambiguity generated in Step (2). The number of syntactic trees to be explored can be exponential. Another known source of complexity for natural language processing is constituted by *anaphoras*, in particular *pronoun* anaphoras, *noun* anaphoras or *elliptic* ones. These are very costly in natural language processing, but, fortunately, we can overlook them when treating legal documents. It is in fact essentially anomalous in legal texts to incorporate references that are not explicit, and therefore in these documents anaphoras are rare, if not inexistent.

Clearly, when dealing with the translation process of legal documents, the fundamental steps to devise depend on the structure of the target formal language. In particular, we aim at devising methods to identify modal operators that represent *obligations*, *permissions* and *prohibitions* and operators that assert *exceptions* to other deontic rules. We also need to detect which tokens can be relevant to the above mentioned deontic operators. Among these tokens we should particularly consider, as usual in syntactic tree generation, nouns and verbs (excluding modal operators that are already considered in the previous step). Moreover, we want to identify *noun phrases*, that are commonly used in place of nouns.

Once the correct pipeline for the process described above is determined, we are able to build an experimental test with human subjects to test the validity of the method; this experiment is still on its way. In order to achieve this long-term goal, first we need to check the correctness of the pipeline on some examples. In this paper we provide a proof-of-concept of the method by implementing the techniques mentioned above to the specific case of contracts and other documents for *natural resources* that are contained in a stable

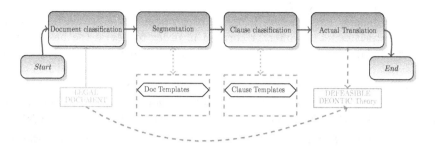

Fig. 1. The pipeline to treat legal texts.

and continuously fed document repository publicly available on the web [1] . We chose one specific case where we have been able to operate the entire pipeline in a semi-automated way, employing the GATE text analysis tools and human analysis. The resulting pipeline is analysed at a high level to identify drawbacks and advantages, and to design the successive step of the process: the experimental phase.

The rest of the paper is organised as follows. Section 2 describes the adopted approach and gives more details about the pipeline. Section 3 shows the application of the method to a specific case: an exploration permit issued to three companies by the government of the State of Western Australia. Section 4 reviews relevant literature, and Section 5 takes some conclusions and sketches further work.

2 Approach

In this section we describe the specific method of natural language processing that we are implementing. The general schema of document classification and knowledge extraction techniques that summarises the existing methods of the current literature is presented in Figure 1.

In the pipeline we introduce the concept of *template* in order to identify (1) patterns of interpretation for legal document categories and (2) segments within one category, that can also be called call *clauses*, as in Figure 1. In the specific case of *exploration permits* that we analyse in this paper, there are essentially only three segments in terms of document template: the introduction or *preamble* segment, where the parts of the permit and its nature are declared; the *interpretation* segment, that contains both *references* to the relevant normative background and *internal definitions* to the relevant terms; and the *prescriptive* segment, where the permittee is assigned with specific duties. If a term does not appear in the interpretation segment because it is not defined nor referenced there, then this term should be treated with a *common sense interpretation*.

When processing a legal document, the performance of the knowledge extraction process is enhanced by knowing *a priori* the category of the document

---

[1] www.resourcecontracts.org

and consequently its structure in terms of segments. We are currently conducting experiments on different corpora in order to devise the best way to extract the knowledge for each corpora. Clearly, it makes sense to identify the properties of the corpora that are useful to the above pipeline, and not those that cannot be used in practice.

The considered approach involves a pipeline that explodes the details of the node **Actual Translation** of Figure 1 and consists of the following steps:

(i) Classify the document in order to identify its *type, language* and *normative background.*

(ii) Apply *named-entity recognition (NER)* to extract entities that are mentioned in the text. This step is carried out by the usage of knowledge extraction and information retrieval methods and aims at extracting:
  • Constitutive elements: definitions of terms that are used in the text.
  • Prescriptive elements: rules that constrain the behaviour of the subjects involved in the legal document, they can be obligations, prohibitions, permissions, or exceptions to those.
  • External references: links to other documents, with possibly a reference to specific terms or rules in these documents.
  • Normative background references: links to the normative background that was in force when the document was drafted.

(iii) *Translate to DDL (defeasible deontic logic) rules* by using the recognized entities. If a translation is available for the other legal texts that are referenced in the current document, then also the corresponding rules are included in the framework.

In order to identify the relevant *verbs, nouns* or *noun phrases* in the text, a *preprocessing step* may be necessary. We will describe a way to do this automatically but at this stage of the research we have automated only the detection, whilst the list of terms to be used in that phase is obtained manually.

In Section 2.1 we specify the formal language of *Defeasible Deontic Logic* [8], that is going to be the technical target of the translation process. Section 2.2 describes the architecture and specifies the functionalities of the well-known GATE system[2], that is employed in this study as a framework for the development of the solution in a prototype.

2.1 Defeasible Deontic Logic

Defeasible logic is a rule-based skeptical approach to non monotonic reasoning. It is based on a logic programming-like language and is a simple, efficient but flexible formalism capable of dealing with many intuitions of non-monotonic reasoning in a natural and meaningful way [1]. Defeasible deontic logic (DDL) is defeasible logic with deontic operators. The formal language that we consider for the translation is a propositional fragment of predicate defeasible deontic logic where no quantified variables appear. Consider a set PROP of proposi-

---

[2] www.gate.ac.uk

tional atoms. The set Lit = PROP $\cup \{\neg p | p \in \text{PROP}\}$ denotes the set of literals. The *complement* of a literal q is denoted by $\sim q$; if q is a positive literal p, then $\sim q$ is $\neg p$, and if q is a negative literal $\neg p$ then $\sim q$ is p.

A defeasible theory D is a tuple $(F, R, >)$. $F \subseteq \text{Lit}$ are the facts, which are always-true pieces of information. R contains three types of rules: strict rules, defeasible rules and defeaters. A rule is an expression of the form $r :$ $A(r) \hookrightarrow C(r)$, where r is the name of the rule, the *arrow* $\hookrightarrow \in \{\rightarrow, \Rightarrow, \rightsquigarrow\}$ is to denote, resp., strict rules, defeasible rules and defeaters, $A(r)$ is the antecedent of the rule, and $C(r)$ is its consequent. A strict rule is a rule in the classical sense: whenever the antecedent holds, so does the conclusion. A defeasible rule is allowed to assert its conclusion unless there is contrary evidence to it. A defeater is a rule that cannot be used to draw any conclusion, but can provide contrary evidence to complementary conclusions. Lastly, $> \subseteq R \times R$ is a binary, antisymmetric relation, with the exact purpose of solving conflicts among rules with opposite conclusions by stating superiorities. We use the following abbreviations on R: R_s is to denote the set of strict rules in R, R_{sd} the set of strict and defeasible rules in R, and $R[q]$ the set of rules in R s.t. $C(r) = q$.

A *derivation* (or *proof*) is a finite sequence $P = P(1), \ldots, P(n)$ of *tagged literals* of the type $+\Delta q$ (q is definitely provable), $-\Delta q$ (q is definitely refuted), $+\partial q$ (q is defeasibly provable) and $-\partial q$ (q is defeasibly refuted). The proof conditions below define the logical meaning of such tagged literals. Given a proof P we use $P(n)$ to denote the n-th element of the sequence, and $P(1..n)$ denotes the first n elements of P. The symbols $+\Delta$, $-\Delta$, $+\partial$, $-\partial$ are called *proof tags*. Given a proof tag $\pm\# \in \{+\Delta, -\Delta, +\partial, -\partial\}$, the notation $D \vdash \pm\#q$ means that there is a proof P in D such that $P(n) = \pm\#q$ for an index n.

In what follows we only present the proof conditions for the positive tags: the negative ones are obtained via the principle of *strong negation*. This is closely related to the function that simplifies a formula by moving all negations to an innermost position in the resulting formula, and replaces the positive tags with the respective negative tags, and the other way around.

The proof conditions for $+\Delta$ describe just forward chaining of strict rules.

$+\Delta$: If $P(n + 1) = +\Delta q$ then either
    (1) $q \in F$, or
    (2) $\exists r \in R_s[q]$ s.t. $\forall a \in A(r)$. $+\Delta a \in P(1..n)$.

Literal q is definitely provable if either (1) it is a fact, or (2) there is a strict rule for q, whose antecedents have all been definitely proved. Literal q is definitely refuted if (1) it is not a fact and (2) every strict rule for q has at least one definitely refuted antecedent.

The conditions to establish a defeasible proof $+\partial$ have a structure similar to arguments in natural language, where an argument might provide support for its conclusion but not be deductively valid in general, because it is defeated by a stronger counter-argument.

$+\partial$: If $P(n+1) = +\partial q$ then either
  (1) $+\Delta q \in P(1..n)$, or
  (2) (2.1) $-\Delta{\sim}q \in P(1..n)$ and
    (2.2) $\exists r \in R_{sd}[q]$ s.t. $\forall a \in A(r) : +\partial a \in P(1..n)$, and
    (2.3) $\forall s \in R[{\sim}q]$. either
      (2.3.1) $\exists b \in A(s) : -\partial b \in P(1..n)$, or
      (2.3.2) $\exists t \in R[q]$ s.t. $\forall c \in A(t) : +\partial c \in P(1..n)$ and $t > s$

A literal q is defeasibly proved if, naturally, it has already strictly proved. Otherwise, we need to use the defeasible part of the theory. Thus, first, the opposite literal cannot be strictly proved (2.1). Then, there must exist an applicable rule supporting such a conclusion, where a rule is applicable when all its antecedents have been proved within the current derivation step. We need to check that all counter-arguments, i.e., rules supporting the opposite, are either discarded (condition (2.3.1), at least one of their premises has been defeasibly rejected), or defeated by a stronger, applicable rule for the conclusion we want to prove (2.3.2).

As in [4], we assume that norms are represented in defeasible deontic logic by the definition that follows:

Definition 2.1 [Norm]
A norm n is a finite set of rules in defeasible deontic logic, where each rule is either a definition $l_1, ..., l_n \to l$, that means a strict rule, a fact l, an unconditional rule with a modal, $\mathcal{M}\, l$, or a conditional rule $l_1, ..., l_n \Rightarrow \mathcal{M}\, l$, where $l, l_1, ..., l_n$, with $n \geq 0$, are propositional literals representing states, actions, or events (asserted to occur or negated as not occurring). $\mathcal{M}$ is a deontic operator indicating an obligation $\mathcal{O}$, a prohibition $\mathcal{F}$, or a negation of one of them.

As common in modal logic, the modals are dualised: in the specific case of deontic logic, given a literal l, $\mathcal{O}l \Leftrightarrow \mathcal{F} \sim l$ and $\mathcal{O} \sim l \Leftrightarrow \mathcal{F}l$.

2.2 Text Processing

GATE is an open-source infrastructure that can be used to develop natural language processing (NLP) software components. It can be used in an interactive manner and it allows to extract information by writing rules that make use of syntactic analyses. We built a custom pipeline with the following language resources:

- *English Tokeniser*, that splits the text into tokens;
- *Gazetteer*, that annotates terms in the text;
- *English Sentence Splitter*, based on punctuation;
- *POS Tagger*, that assigns POS (Part of Speech) tags to tokens;
- *JAPE Transducer*, that attempts to find unknown Named Entities based on extraction templates written in the JAPE language.

A *gazetteer list* is a pre-made lookup list adopted to annotate terms in the text and perform Named Entity Recognition. More complex formulae built with these terms can be detected by using JAPE (Java Annotation Pattern Engine), that allows to specify regular expressions that make use of those simpler annotations. For instance, it can be used to annotate as an NE of type Person lookups of type "title" followed by a "firstname" and "lastname" or to annotate as an NE of type Organization lookups with an NNP (proper noun) POS tag followed by an annotation for Company suffixes ("Ltd" or "GmBH").

In order to perform Named Entity Recognition, a list of relevant words or locutions have to be fed to the gazetteer resource. Performing this step in an automated step could be potentially disruptive, as shall be clear in the application of the pipeline to one sample document in Section 3, and therefore we are planning a new experiment with this specific purpose.

3 Case Study: an Exploration Permit

In this section we apply the methodology described in the previous section to a real-world resource contract. We tried our approach on contracts from the *resourcecontracts.org* website, that contains thousands of petroleum and mining contracts, and chose to present the analysis of one specific contract [3] with the features that we are interested in, but that also maintains a small size, so that it is readable and understandable by humans. The contract concerns the grant of an exploration permit for petroleum in the State of Western Australia to three companies. The analysis is carried out on the .doc version of the document, containing 1264 words.

We apply the pipeline to the document as follows. First, we apply the preprocessing step, where we detect the relevant words or locutions in the text. Then, we execute the first step of the pipeline to classify the document in order to identify its type, language and normative background. In our case, the type is inferred by the title, the English language is contained in the metadata of the repository, and the normative background is also inferred by the title, where "State of Western Australia" appears. Moreover, there is a reference to the Petroleum Act (1967) in the first page and a reference to the Aboriginal Heritage Act (1972) in the last page; these references are found later by a JAPE rule.

Subsequently, we apply named-entity recognition (NER) to extract named entities that are mentioned in the text. The Named Entity Recognition process detects Acts (Petroleum Act, 1967 and Aboriginal Heritage Act, 1972), Locations (State of Western Australia, San Francisco, California, USA, Tulsa, Oaklahoma, USA), Organizations (GEOPETRO COMPANY, SEVEN SEAS AUSTRALIA INC., AMITY OIL NL), Persons (NORMAN MOORE, Minister for Mines, inspector, WILLIAM LEE TINAPPLE, the permittee, Director Petroleum Operations Division, a person), and Dates (JULY 2, 1997, June 1998, November 1998).

---

[3] The contract that we analyse can be found at https://www.resourcecontracts.org/contract/ocds-591adf-5231394526/view.

Gazetteer lists contain pre-defined terms that we are interested to detect in the text, while more complex JAPE rules are used to detect the combination of a prescription with a verb and an object. We give more details about the process while focusing on "Schedule 2" of the document, that is the section that contains prescriptive elements. In Figure 2 we show the LHS of the JAPE rules that are adopted to detect obligations, permissions, prohibitions and exceptions: these rules detect patterns containing terms that appear in gazetteer lists (i.e., `obligation`, `permission`, `prohibition`, and `exception`) and terms tagged as specific parts of speech (such as verbs VB or nouns NN), and tag them as relations (i.e., `ObligationRelation1`, `PermissionRelation1`, `ProhibitionRelation1`, and `ExceptionRelation1`).

Deontic modalities are identified by the tokens **shall** and **must** for obligations (translated to the operator $\mathcal{O}$); the expression **may** for permissions (translated to the operator $\neg\mathcal{F}$); **shall not** for prohibitions (translated to the operator $\mathcal{F}$); and **except** for exceptions (translated to a precedence > between rules). On the other hand, named entities that are recognized as Locations, Organizations, Persons, and Dates will be translated to constant terms in DDL.

```
(({obligation})
({Token.category==VB})
({Token.category!=NN})*
({Token.category==NN})
):ObligationRelation1

(({prohibition})
({Token.category==VB})
({Token.category!=NN})*
({Token.category==NN})
):ProhibitionRelation1
```

```
(({permission})
({Token.category==VB})
({Token.category!=NN})*
({Token.category==NN})
):PermissionRelation1



(({exception})
({Token.category==VB})
):ExceptionRelation1
```

Fig. 2. JAPE rules to detect obligations, permissions, prohibitions and exceptions

Some of the expressions that appear in the text deserve specific attention as they can be classified as specific *legal terms*, and therefore can be easily bound with specific gazetteers, which is more convenient than to apply general rules. The number of expressions that need to be captured in this manner is rather low in the practice of legal text processing, and these special cases are usually just verbs or verbal locutions. Once a specific subdomain has been engineered in terms of document corpus for this goal, the objects inserted in the gazetteers shall be limited to a controlled number, and therefore manageable in practice.

In the specific case of the exploration permit that we applied the techniques to, we have identified on purpose some cases that need this special treatment. The special terms we encountered are: *in accordance with the approval, production testing, measuring, permit...to test, authorised in writing for the purpose, take adequate measures, comply.*

The expression *in accordance with the approval* is treated as special because the structure of the corresponding noun phrase is inserted in a specific context, that of providing an exception to another rule, that occurs in many legal texts for the specific noun *approval*. The expressions *production testing* and *measuring* are in the gerund form, that is somewhat uncommon in many other cases but can appear in legal texts. A good approach to solve this is purely *syntactic* as the majority of verbs form regular gerunds. The expression *permit...to test* contains an expression (permit)~ that could be translated into a modal, but for the purpose of this investigation we limited ourselves to modals whose scope is on a propositional structure.

The expression *authorised in writing for the purpose* is a relative clause, and we take *authorised* as the corresponding past passive expression. Again, we detect this by having it in the gazetteer list. The idiosyncratic expression *take adequate measures* is common in technical language related to legal texts. The expression *comply* could be considered potentially redundant in terms of application. The full expression contains a basic obligation: the permittee has to comply with the requests of the Minister for Mines. This is in some sense intrinsic in the definition of an exploration permit for natural resources, but it is also asserted in the referenced Petroleum Act for what concerns exploration permits.

In Table 1 and 2 we enumerated the rules by an automated mechanism that generates the rule labels by the enumeration found in the text. When a single piece of text is translated into more than one single rule, the enumeration builds a # pattern at the end of the automatically generated string and follows it with a counter that starts from 1. The superiority relation > is established between the rules as a direct effect of the exception operators found in the text: $r1.2\#2 > r1.2\#1$, $r2\#2 > r2\#1$, $r4\#2 > r4\#1$, and $r5\#2 > r5\#1$.

Finally, the recognized entities are used to create DDL (defeasible deontic logic) rules. The JAPE rules allow to tag complex expressions, but this step is still performed mainly manually in the majority of applications. We have developed this part of the pipeline by adding elements by hand and processing them in an automated way. The process can be represented by the following meta-rules for the JAPE rules in Figure 2:

- In the case of *ObligationRelation1*, *PermissionRelation1* or *ProhibitionRelation1*, represented by a tuple $< M, V, NE_1, \ldots, NE_n >$, where M is a modality , V is a term of type predicate, and NE_i $(1 \leq i \leq n)$ is a term of type constant, we translate to a rule $\Rightarrow \mathcal{M} \; p(ne_1, \ldots, ne_n)$, where $\mathcal{M}$ is the modal operator corresponding to M, p is the predicate corresponding to V and ne_i is the constant corresponding to NE_i;

- In the case of *ExceptionRelation1*, represented by a tuple $< E, Q >$, where E is an exception and Q is a term of type predicate, that follows a tuple $< M, V, NE_1, \ldots, NE_n >$, defined and translated as above, we translate to a rule $q(ne_1, \ldots, ne_n) \Rightarrow \sim \mathcal{M} \; p(ne_1, \ldots, ne_n)$, where q is the predicate corresponding to Q and this rule has precedence over the previous rule.

Tables 1 and 2 show the final translation of "Schedule 2" of the document in DDL. Table 1 shows original text and DDL formulae for the first part of the

Table 1
Translation to DDL: constants , modalities , predicates , exceptions .

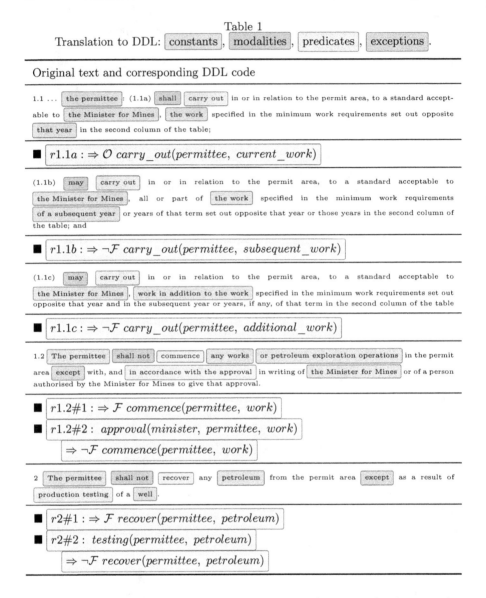

Original text and corresponding DDL code

1.1 ... the permittee : (1.1a) shall carry out in or in relation to the permit area, to a standard acceptable to the Minister for Mines , the work specified in the minimum work requirements set out opposite that year in the second column of the table;

■ $r1.1a : \Rightarrow \mathcal{O}\ carry\_out(permittee,\ current\_work)$

(1.1b) may carry out in or in relation to the permit area, to a standard acceptable to the Minister for Mines , all or part of the work specified in the minimum work requirements of a subsequent year or years of that term set out opposite that year or those years in the second column of the table; and

■ $r1.1b : \Rightarrow \neg\mathcal{F}\ carry\_out(permittee,\ subsequent\_work)$

(1.1c) may carry out in or in relation to the permit area, to a standard acceptable to the Minister for Mines , work in addition to the work specified in the minimum work requirements set out opposite that year and in the subsequent year or years, if any, of that term in the second column of the table

■ $r1.1c : \Rightarrow \neg\mathcal{F}\ carry\_out(permittee,\ additional\_work)$

1.2 The permittee shall not commence any works or petroleum exploration operations in the permit area except with, and in accordance with the approval in writing of the Minister for Mines or of a person authorised by the Minister for Mines to give that approval.

■ $r1.2\#1 : \Rightarrow \mathcal{F}\ commence(permittee,\ work)$

■ $r1.2\#2 : approval(minister,\ permittee,\ work)$
  $\Rightarrow \neg\mathcal{F}\ commence(permittee,\ work)$

2 The permittee shall not recover any petroleum from the permit area except as a result of production testing of a well .

■ $r2\#1 : \Rightarrow \mathcal{F}\ recover(permittee,\ petroleum)$

■ $r2\#2 : testing(permittee,\ petroleum)$
  $\Rightarrow \neg\mathcal{F}\ recover(permittee,\ petroleum)$

text, while for the rest Table 2 shows only the DDL formulae. At this step also the logical rules of the two referenced documents Petroleum Act (1967) and Aboriginal Heritage Act (1972) are included in the logical theory.

4 Related works

Computational methods applied to the law can be distinguished in two general approaches (see [5]): the *law-as-code* approach and the *law-as-data* approach. The law-as-code approach aims at interpreting and representing legal rules in a formal language, such as defeasible deontic logic [8]. An example is that of

Table 2
Translation to DDL.

DDL code
■ $r3a : \Rightarrow \mathcal{O}\ pay(permittee,\ minister,\ petroleum)$
■ $r3b : \Rightarrow \mathcal{O}\ furnish(permittee,\ minister,\ petroleum,\ particulars)$
■ $r3c :\ authorised(person),\ measuring(device,\ petroleum)$ $\Rightarrow \mathcal{O}\ permit\_test(permittee,\ person,\ device)$
■ $r4\#1 : \Rightarrow \mathcal{F}\ install(permittee,\ equipment$ ■ $r4\#2 :\ approval(minister,\ permittee,\ equipment)$ $\Rightarrow \neg\mathcal{F}\ install(permittee,\ equipment)$
■ $r5\#1 : \Rightarrow \mathcal{F}\ abandon(permittee,\ well)$ ■ $r5\#2 :\ approval(minister,\ permittee,\ well)$ $\Rightarrow \neg\mathcal{F}\ abandon(permittee,\ well)$

[2], where the authors provide a method for encoding traffic regulatory rules.

On the other hand, the law-as-data approach aims at extracting information from high-dimensional legal datasets. This approach can be applied to many problems in the legal context, and applications can range from interpretation of legal texts to quantitative analysis of external factors that influence the law. In this regard, there exist different pattern-based approaches to assign labels to (parts of) the text of legal documents. For instance in [9] authors perform automatic categorization of case law documents into 40 high-level categories. The system Salomon [11] processes Belgian criminal cases by performing an initial categorisation and structuring of the texts and an extraction of the most relevant text units. On the other hand, in [6] linguistic information such as lemmatisation and part-of-speech tags is used to improve the classification of Portuguese legal texts. Work in [7] combines learning and reasoning to automatically detect and explain unfair clauses in Terms of Services of online consumer contracts. Authors in [10] propose a method for classifying the sub-components of the writing styles that bound German legal language, and a specialized annotated corpus is again necessary.

The work that we described in this article sits in the middle between the two approaches: we start from a dataset of natural-language contracts that is interpreted and synthesised to rules in defeasible deontic logic.

5 Conclusions and further investigations

In this paper we discussed the development of a pipeline for translating legal texts into a formal language, defeasible deontic logic. The methodology we developed is then applied to a sample case, an exploration permit for petroleum in Western Australia. The permit is part of the publicly available *resourcecontracts.org* corpus, a repository of more than 2500 legal documents related to natural resources.

The paper presents a proof-of-concept for the above mentioned pipeline that is going to be employed to process the documents in this corpus in order to identify varieties of the pipeline behaviour and determine which methods better fit the translation pipeline. The choice of the mentioned corpus is fourfold, as each of its corpus elements is (1) coherent from the viewpoint of the matters it discusses; (2) not related to a specific normative background; (3) multi-lingual (sometimes also the same document is written in more than one language); and (4) diverse in terms of forms.

So far, the results we obtained are very promising, as the number of basic errors is less than one might expect and the quality of the translation is rather good, although only partial at this stage. We are also carrying out a gold standard test, where a group of legal experts shall check the validity of the translation with a specific methodology by comparing the application of the automated method with their legal reasoning to measure the difference.

References

[1] Antoniou, G., D. Billington, G. Governatori and M. J. Maher, *Representation results for defeasible logic*, ACM Transactions on Computational Logic **2** (2001), pp. 255–287.

[2] Bhuiyan, H., F. Olivieri, G. Governatori, M. Islam, A. Bond and A. Rakotonirainy, *A methodology for encoding regulatory rules*, CEUR Workshop Proceedings **2632** (2020).

[3] Camilleri, J., N. Gruzitis and G. Schneider, *Extracting formal models from normative texts*, Lecture Notes in Computer Science (including subseries Lecture Notes in Artificial Intelligence and Lecture Notes in Bioinformatics) **9612** (2016), pp. 403–408.

[4] Cristani, M., C. Tomazzoli, F. Olivieri and L. Pasetto, *An ontology of changes in normative systems from an agentive viewpoint*, Communications in Computer and Information Science **1233 CCIS** (2020), pp. 131–142.

[5] Frankenreiter, J. and M. Livermore, *Computational methods in legal analysis*, Annual Review of Law and Social Science **16** (2020), pp. 39–57.

[6] Gonçalves, T. and P. Quaresma, *Is linguistic information relevant for the classification of legal texts?*, Proceedings of the International Conference on Artificial Intelligence and Law (2005), pp. 168–176.

[7] Lagioia, F., F. Ruggeri, K. Drazewski, M. Lippi, H.-W. Micklitz, P. Torroni and G. Sartor, *Deep learning for detecting and explaining unfairness in consumer contracts*, Frontiers in Artificial Intelligence and Applications **322** (2019), pp. 43–52.

[8] Nute, D., "Defeasible Deontic Logic," Springer Publishing Company, Incorporated, 2010, 1st edition.

[9] Thompson, P., *Automatic categorization of case law*, Proceedings of the International Conference on Artificial Intelligence and Law (2001), pp. 70–77.

[10] Urchs, S., J. Mitrovic and M. Granitzer, *Towards classifying parts of german legal writing styles in german legal judgments*, 2020 10th International Conference on Advanced Computer Information Technologies, ACIT 2020 - Proceedings (2020), pp. 451–454.

[11] Uyttendaele, C., M.-F. Moens and J. Dumortier, *Salomon: Automatic abstracting of legal cases for effective access to court decisions*, Artificial Intelligence and Law **6** (1998), pp. 59–79.

[12] Wyner, A. and W. Peters, *On rule extraction from regulations*, Frontiers in Artificial Intelligence and Applications **235** (2011), pp. 113–122.

Arguing coalitions in abstract argumentation

Lisha Qiao[1,2]   Yiqi Shen[4]   Liuwen Yu[1,2,3]   Beishui Liao[1,4]
Leendert van der Torre[1,4]

[1]University of Luxembourg, Luxembourg
[2]University of Bologna, Italy
[3]University of Turin, Italy
[4]Zhejiang University, China

Abstract

In this paper, we are interested in different ways in which agents can collaborate in abstract agent argumentation. First, if arguments are accepted when they are put forward by more than one agent, then agents can put forward arguments from other agents of the coalition. Second, agents can put forward arguments to defend arguments from other agents of the coalition. For example, in expert opinion, a domain expert can put forward an argument defending an argument made by a politician, even when the politician cannot judge the correctness of the argument. Third, agents from a coalition can collectively defend an argument they share, without being able to defend the argument individually. In this paper, we formalize the different kinds of collaboration in abstract agent argumentation, and we illustrate the coalition formation with a case study in political debate.

Keywords: abstract argumentation, abstract agent argumentation, coalition formation

1 Introduction

Dung [14] defines an argumentation framework as a set of arguments and a binary relation between them. Agent argumentation frameworks [24] extend Dung's theory with agents or the sources of the arguments. Most approaches to agent argumentation are inspired by social choice and voting theory, and prefer arguments or attacks if they belong to more than one agent [5,6,10,17,20]. Moreover, in this setting, several authors [1,2,3,18] have considered the role of coalitions, for example in the setting of coalition formation [2]. Building on this line of work, in this paper we are interested in the following research question:

How to study coalition formation in abstract agent argumentation?

To answer this question, we adopt the minimal formal framework we introduced earlier for our principle-based analysis of agent argumentation semantics [24]. Agent-based extensions typically introduce various aspects such as knowledge, uncertainty, support, trust and so on. We use a minimal extension of Dung's

theory [14] as a common core in those approaches. We limit this paper to an
abstract set of agents, associate arguments with agents, and a partitioning of
the set of agents to represent coalitions. Our research question breaks down
into the following sub-questions:

(1) Which kinds of collaboration among arguing agents can be distinguished?

(2) How can Dung's notion of defense [14] be adapted to incorporate coalitions?

(3) How can we use this theory to reason about coalition formation?

To answer the first sub-question, we distinguish between three reasons for
forming coalitions:

(1) Arguments can be accepted when they are put forward by more than one
agent.

(2) Arguments can be defended by arguments from other agents.

(3) An argument shared by several agents can be collectively defended by a
group without being defended by an individual agent.

For example, in expert opinion, a domain expert can put forward an argument
that defends an argument made by a politician, even when the politician cannot
judge the correctness of the argument.

To answer the second sub-question, we introduce the notion of coalition
defense: arguments can only be defended by arguments from the same coalition.
Everything else stays the same.

To answer the third sub-question, we use a running example from coalition
formation in politics. A libertarian, a collectivist, an anarchist and a political
expert form different coalitions in order to have as many of their arguments
accepted as possible.

The potential application of this paper is to see how we can maximize
agreement when we want to form a coalition in multi-agent systems (MASs). In
the real world, we can also use this coalition framework to distinguish between
different reasons for forming coalitions. Our investigation highlights various
properties which can be further studied in this formal setting.

The layout of this paper is as follows. Section 2 introduces a case study in
politics, and we informally describe the different kinds of collaboration in the
example. Section 3 introduces the notion of coalition argumentation framework. Section 4 formalizes coalition defense as well as coalition semantics, and
uses the running example to further explain these notions. Section 5 discusses
several properties. We discuss related work in section 6, and future work in
section 7. Section 8 concludes the paper.

2   Coalition formation in politics

In this section, we introduce a running example from political debate.

2.1 Political views

Let us suppose that there are four kinds of people with different political tendencies: Libertarian (L, for short), Collectivist (O, for short), Anarchist (N, for short), and Political expert(E, for short). They hold different points of view on whether government and laws are necessary for society.

L's point of view involves improving the functions of government and the law. We need government and laws. What government and laws do first of all is protect individual freedom.

O holds that we should improve the functions of government and law. Our government and laws protect the collective interest, so everyone can gain more interest. Sometimes, we need to limit individual freedom to comply with the law.

N argues that we don't need any government or law to protect individual freedom. We only need a government that serves people's interests. The government and laws that serves the elite firstly protect the individual interests of that minority. So, we don't need the current government and law.

E has two arguments: 1) today's elite-led government is committed to realizing collective interests, and 2) we need to limit individual freedom to comply with the law.

2.2 Political conflict

Obviously, there are conflicts in the views of these politicians. Even one politician's views will be self-contradictory.

When N only chooses to accept one of his/her own ideas, N prefers the former viewpoint that we don't need any government or law to protect individual freedom because even good government and laws can protect individual interests, including freedom; they may also limit and even deprive individuals of their personal freedom.

L advocates that we need government and laws whose first aim is to protect individual freedom, and objects to N's view that we don't need any government or law. L's point is more robust than N's because government and law are historical choices. And L's argument also contradicts E's argument that we need to limit individual freedom to comply with the law because individual freedom is not absolute but relative; if we obey the law, we don't need to restrict our freedom.

E's argument also attacks L's view when recognizing that freedom is absolute and needs to be limited. The conflict between N and L is clear. We cannot accept both L's view and N's view. N's two points are both stronger than L's view. Because we regard freedom as the most important thing, and any government and law can limit our freedom, we accept N's argument.

A conflict between N and O is also apparent. O is of the same opinion as L that we should improve the functions of government and the law. We already know that N's argument is more robust from this viewpoint. When it comes to individual interests, N's view is that the current government and law prioritize

protecting the interests of minority people (the elite), and that we don't need
the present government and law. O holds the opposite view.

Intuitively, political experts have professional political knowledge, and their
viewpoints are often accepted. But N strongly contradicts the experts' statement. We cannot deny the existence of elite interest groups who are committed
to realizing the interests of minority group members firstly. Political experts
also admit this. So, N's proposal that we don't need the current government
or law proposed is accepted.

2.3   Political coalition structure

There are contradictions in the views of every politician, and to make their
ideas accepted, they can choose whether or not to form a coalition. These four
politicians can create 15 coalition structures through partitioning. They can
work in a grand coalition or work independently. Their coalitions can also be
two against two (2:2), two against one against one (2:1:1), and three against
one(3:1). For example, L and N form a coalition, O and E form another coalition. Alternatively, L, O and N form a coalition, and E works independently.

2.4   Working in a political coalition

If they can work in a grand coalition, O and E both support the view that we
need to limit individual freedom to comply with the law. Intuitively, the more
politicians who support this point of view, the more this view will be accepted.
So, it is possible that the member of this coalition can accept this point of view
in a coalition.

E has two arguments, and these arguments are attacked by L and N respectively. If (s)he doesn't collaborate with other politicians, (s)he cannot let other
politicians accept his/her perspective. Suppose (s)he collaborates with other
politicians to help argue that the current government and laws give priority
to protecting the personal interests of the minority (the elite), and we don't
need our current government and laws. In that case, his/her view that the
current government is committed to realizing collective interests is likely to be
accepted. Thus, E can collaborate with O or N. Although N attacks E, another
view by N can help E refute N. Without the coalition, E does not know whose
view can help him/her refute N.

L and O have a shared view that we should improve government and legal
functions, but N has two arguments for attacking this view. If they work in
the same coalition, they know that they can attack N's two points of view
respectively such that their shared view can be accepted.

3   Coalition argumentation framework

This section introduces coalition argumentation frameworks. Coalition argumentation frameworks generalize argumentation frameworks studied by
Dung [14], which are directed graphs, where the nodes are arguments, and
the arrows correspond to the attack relation.

A coalition argumentation framework extends an argumentation framework with a set of agents, a relation associating arguments with agents, and a partitioning of the agents called a coalition structure. An argument can belong to no agent, one agent or multiple agents [24]. Each agent belongs to exactly one coalition.

We write $a \sqsubset \alpha$ to represent that argument a belongs to agent α, or that agent α has argument a. We write $C = \{\alpha\beta\}$ if agents α and β are in the same coalition, e.g. they work together in a political coalition.

Definition 3.1 [Coalition argumentation framework] A *coalition argumentation framework* (CAF) is a 5-tuple $\langle \mathcal{A}, \rightarrow, \mathcal{S}, \sqsubset, C \rangle$ where $\mathcal{A}$ is a set of arguments, $\rightarrow \subseteq \mathcal{A} \times \mathcal{A}$ is a binary relation over $\mathcal{A}$ called attack, $\mathcal{S}$ is a set of agents or sources, $\sqsubset \subseteq \mathcal{A} \times \mathcal{S}$ is a binary relation associating arguments with agents, and $C = \{\mathcal{S}_1, \mathcal{S}_2, ... \mathcal{S}_n\}$ is a set of disjoint subsets of $\mathcal{S}$ such that every agent of $\mathcal{S}$ occurs in exactly one of these subsets, representing the coalition structure.

The following example illustrates the formalization of the running example as a coalition argumentation framework.

Example 3.2 [Coalitions] Consider the running example in section 2. There are four agents L, E, O and N. Their arguments are as follows.
L has two arguments i and a.
    i: We should improve the functions of government and the law.
    a: We need government and laws.

O has three arguments i, g, l.
    i: We should improve the functions of government and the law.
    g: Our government and laws guarantee the collective interest, so everyone can gain more interest.
    l: Sometimes, we need to limit individual freedom to comply with the law.

N has two arguments n and o.
    n: We don't need any government and laws to protect individual freedom.
    o: We only need the government and laws that serve people's interests.

E has two arguments c and l.
    c: Today's elite-led government is committed to the collective interest.
    l: We need to limit individual freedom to comply with the law.

Consider the coalition argumentation framework of the running example depicted in Figure 1, which contains $\mathcal{A} = \{i, n, o, a, g, c, l\}$, $\rightarrow = \{a \rightarrow n, n \rightarrow i, n \rightarrow o, g \rightarrow o, o \rightarrow g, o \rightarrow i, o \rightarrow c, l \rightarrow a, a \rightarrow l\}$, $\mathcal{S} = \{$L,E,O,N,$\}$, $\sqsubset = \{(a, L), (i, L), (n, N), (o, N), (i, O), (g, O), (c, E), (l, O), (l, E)\}$.

An example of a coalition structure is $C = \{$LON,E$\}$, in which L, O and N work together and E works independently. We abbreviate this to $C = \{$LON,E$\}$. Using this abbreviation, all possible coalition structures are as follows. $C_1 = \{$LEON$\}$, $C_2 = \{$L,O,N,E$\}$, $C_3 = \{$LON,E$\}$, $C_4 = \{$LEO,N$\}$, $C_5 = \{$ONE,L$\}$, $C_6 = \{$LEN,O$\}$, $C_7 = \{$LO,NE$\}$, $C_8 = \{$LN,OE$\}$, $C_9 = \{$ON,LE$\}$, $C_{10} = \{$LO,N,E$\}$, $C_{11} = \{$LN,O,E$\}$, $C_{12} = \{$LE,O,N$\}$, $C_{13} = \{$ON,L,E$\}$, $C_{14} = \{$OE,L,N$\}$, $C_{15} = \{$NE,L,O$\}$.

Note that in this example, all arguments belong to at least one agent, and arguments i and l belong to two agents. Moreover, note that the arguments of most agents do not conflict with their own other arguments, only with those of the other agents. An exception is arguments n and o which belong to agent N. The attack from n to o means that without any interaction with other agents, agent N accepts n, but if this argument is rejected due to attacks from other agents, (s)he accepts argument o.

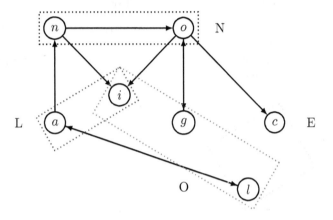

Fig. 1. Abstract framework of the political debate without coalitions

4   Coalition defense semantics

We now introduce a new kind of defense for coalition argumentation frameworks, which we call coalition defense. We adapt Dung's notions of defense and admissibility for coalition argumentation frameworks. Dung [14] defines conflict-freeness as a situation where attacking and attacked arguments cannot be accepted at the same time. A set of arguments defends an argument when the former attacks all attackers of the latter. A set is admissible if it is conflict-free and it defends all its elements. In our theory, roughly, if an agent puts forward an argument, it can only be coalition defended by arguments belonging to agents from the same coalition. Coalition admissibility means that a set of arguments is conflict-free and the coalition defends all its elements.

- **Definition 4.1** [Coalition admissible] Let $\langle \mathcal{A}, \rightarrow, \mathcal{S}, \sqsubset, C \rangle$:
  - $E \subseteq \mathcal{A}$ is *conflict-free* iff there are no arguments a and b in E such that a attacks b.
  - $E \subseteq \mathcal{A}$ coalition defends c iff there is a group of agents G in C, there exists an agent α in G having argument c, and for all arguments b in $\mathcal{A}$ attacking c, there exists an argument a in E such that a attacks b, and there is an agent β in G having a.
  - $E \subseteq \mathcal{A}$ is *coalition admissible* iff it is conflict-free and a coalition defends all its elements.

Example 4.2 [Political coalition, continued from Example 3.2] Consider the running example depicted in Figure 1. It illustrates several arguments that can be defended by arguments from other agents in the same coalition. For example, g defends c iff O and E are in the same coalition, and l defends n iff O and N are in the same coalition. Note also that n defends g iff N and O are in the same coalition, even when g also defends itself.

Figure 1 also illustrates that an argument shared by several arguments can be attacked by several arguments, where each argument cannot be defended against all these attacks by an individual agent, but can be defended when the agents work together. For example, a and g defend i iff L and O are in the same coalition.

Dung [14] presents admissibility-based semantics, which refers to an evaluation standard for the acceptability of argument sets and is used to select an acceptable argument set from a group of conflicting arguments. Coalition semantics is an evaluation criterion for the acceptability of argument sets, which is used to select acceptable argument sets from a set of contradictory arguments. It is represented as coalition extensions which are sets of acceptable arguments. And an empty set is considered to defend itself. The difference between coalition semantics and Dung's semantics is that in coalition semantics, arguments can only be defended by arguments from the coalition.

Each extension can be regarded as a set of collectively acceptable arguments. There are four kinds of coalition extensions: coalition complete extensions, coalition grounded extensions, coalition preferred extensions and coalition stable extensions. A coalition complete extension is conflict-free, and it can defend all elements and contains all the arguments it defends, so it's a set of closure. Coalition grounded extensions are the smallest coalition complete extensions, which minimizes the set of compatible arguments. Coalition preferred extensions are the largest coalition complete extensions, which maximizes the set of compatible arguments. Coalition stable extensions are conflict-free and attack all arguments that do not belong to the coalition.

Definition 4.3 [Coalition extensions] Let $\langle \mathcal{A}, \rightarrow, \mathcal{S}, \sqsubset, C \rangle$:

- $E \subseteq \mathcal{A}$ is a *coalition complete extension* iff E is coalition admissible and it contains all the arguments the coalition defends, $E = \{a | E \text{ coalition defends a}\}$.

- $E \subseteq \mathcal{A}$ is a *coalition grounded extension* iff it is the smallest (for set inclusion) coalition complete extension.

- $E \subseteq \mathcal{A}$ is a *coalition preferred extension* iff it is the largest (for set inclusion) coalition complete extension.

- $E \subseteq \mathcal{A}$ is a *coalition stable extension* iff it is conflict-free and it attacks all the arguments in $\mathcal{A}\backslash E$.

The following example illustrates coalition extensions.

Example 4.4 [Political coalition, continued from Example 4.2] Reconsidering the running example depicted in Figure 1, the extensions of each coalition are

Table 1

The coalition extension of all possible coalition structures. We abbreviate Coalition Grounded as $\mathbb{CG}$.

Sem.	Coalition Complete	CG	Coalition Preferred	Coalition Stable
$C_1 = \{\text{LEON}\}$	$\{\emptyset, \{a\}, \{g,c\}, \{a,o\}, \{a,g,c,i\}, \{l,n,g,c\}\}$	$\{\emptyset\}$	$\{\{a,o\}, \{a,g,c,i\}, \{l,n,g,c\}\}$	$\{\{a,o\}, \{a,g,c,i\}, \{l,n,g,c\}\}$
$C_2 = \{\text{L,E,N,O}\}$	$\{\emptyset, \{a\}, \{g\}, \{l\}, \{a,g\}, \{l,g\}\}$	$\{\emptyset\}$	$\{\{a,g\}, \{l,g\}\}$	$\times$
$C_3 = \{\text{LNO,E}\}$	$\{\emptyset, \{a\}, \{g\}, \{a,o\}, \{a,g,i\}, \{l,n,g\}\}$	$\{\emptyset\}$	$\{\{a,o\}, \{a,g,i\}, \{l,n,g\}\}$	$\{\{a,o\}\}$
$C_4 = \{\text{LEO,N}\}$	$\{\emptyset, \{a\}, \{l\}, \{g,c\}, \{l,g,c\}, \{a,g,c,i\}\}$	$\{\emptyset\}$	$\{\{l,g,c\}, \{a,g,c,i\}\}$	$\{\{a,g,c,i\}\}$
$C_5 = \{\text{ENO,L}\}$	$\{\emptyset, \{a\}, \{g,c\}, \{a,g,c\}, \{l,n,g,c\}\}$	$\{\emptyset\}$	$\{\{a,g,c\}, \{l,n,g,c\}\}$	$\{\{l,n,g,c\}\}$
$C_6 = \{\text{ELN,O}\}$	$\{\emptyset, \{a\}, \{l\}, \{g\}, \{a,o\}, \{a,g\}, \{l,g\}\}$	$\{\emptyset\}$	$\{\{a,o\}, \{a,g\}, \{l,g\}\}$	$\{\{a,o\}\}$
$C_7 = \{\text{LO,EN}\}$	$\{\emptyset, \{a\}, \{l\}, \{g\}, \{l,g\}, \{a,g,i\}\}$	$\{\emptyset\}$	$\{\{l,g\}, \{a,g,i\}\}$	$\times$
$C_8 = \{\text{LN,EO}\}$	$\{\emptyset, \{a\}, \{l\}, \{g,c\}, \{a,o\}, \{a,g,c\}, \{l,g,c\}\}$	$\{\emptyset\}$	$\{\{a,o\}, \{a,g,c\}, \{l,g,c\}\}$	$\{\{a,o\}\}$
$C_9 = \{\text{NO,EL}\}$	$\{\emptyset, \{a\}, \{g\}, \{l,n\}, \{a,g\}, \{l,n,g\}\}$	$\{\emptyset\}$	$\{\{a,g\}, \{l,n,g\}\}$	$\times$
$C_{10} = \{\text{LO,E,N}\}$	$\{\emptyset, \{a\}, \{l\}, \{g\}, \{l,g\}, \{a,g,i\}\}$	$\{\emptyset\}$	$\{\{l,g\}, \{a,g,i\}\}$	$\times$
$C_{11} = \{\text{LN,E,O}\}$	$\{\emptyset, \{a\}, \{g\}, \{l\}, \{a,o\}, \{a,g\}, \{l,g\}\}$	$\{\emptyset\}$	$\{\{a,o\}, \{a,g\}, \{l,g\}\}$	$\{\{a,o\}\}$
$C_{12} = \{\text{EL,N,O}\}$	$\{\emptyset, \{a\}, \{l\}, \{g\}, \{a,g\}, \{l,g\}\}$	$\{\emptyset\}$	$\{\{a,g\}, \{l,g\}\}$	$\times$
$C_{13} = \{\text{ON,E,L}\}$	$\{\emptyset, \{a\}, \{g\}, \{l,n\}, \{a,g\}, \{l,n,g\}\}$	$\{\emptyset\}$	$\{\{a,g\}, \{l,n,g\}\}$	$\times$
$C_{14} = \{\text{EO,L,N}\}$	$\{\emptyset, \{a\}, \{l\}, \{g,c\}, \{a,g,c\}, \{l,g,c\}\}$	$\{\emptyset\}$	$\{\{a,g,c\}, \{l,g,c\}\}$	$\times$
$C_{15} = \{\text{EN,L,O}\}$	$\{\emptyset, \{a\}, \{l\}, \{g\}, \{a,g\}, \{l,g\}\}$	$\{\emptyset\}$	$\{\{a,g\}, \{l,g\}\}$	$\times$

listed in Table 1.

If all agents are in a grand coalition, it is the same as Dung's graph. If all agents work independently, a and g, or l and g can be accepted. a and g cannot coalition defend i, because these arguments do not pertain to agents from the same coalition. i is not accepted. i is accepted when L and O are in the same coalition, for example $C_3 = \{\text{LNO,E}\}$. i is not accepted when L and O are not in the same coalition, such as $C_2 = \{\text{L,E,N,O}\}$. An argument shared by several agents can be attacked by several arguments, where each argument cannot be defended against all these attacks by an individual agent, but can be defended when the agents work together.

c is not accepted when E and O are not in the same coalition, like in $C_3 = \{\text{LNO,E}\}$. We accept c when O and E are in the same coalition or N and E are in the same coalition, such as $C_6 = \{\text{LEN,O}\}$. We can say that arguments can be defended by arguments from other agents in the same coalition.

Example 4.5 [Political coalition, continued from Example 4.4] Reconsider the running example depicted in Figure 1. We now briefly explain the idea of social semantics that prefers arguments that belong to more than one agent. Since argument l belongs to two agents and argument a belongs to only one agent, we can say that argument l is preferred to argument a. This can be represented by removing the attack from a to l. The effect on the extensions in Table 1 is that l is now in all extensions, and all extensions with a can be removed from Table 1.

Definition 4.6 [Restricted coalition defend] Let $\langle \mathcal{A}, \rightarrow, \mathcal{S}, \sqsubset, C \rangle$:

E restricted coalition defends $a \iff$

- E coalition defends a (e.g. based on coalition C_i)

- $\forall$ agent $\alpha \in C_i$, $\forall$ argument $\beta \sqsubset \alpha$, a and β do not attack each other.

This definition makes sure that a coalition cannot be formed randomly. Agents who hold contradictory arguments will not tend to form coalitions in most cases, except when they have no conflicting interests. By this definition, agents who holds contradictory arguments can still form a coalition, but the cost is that they cannot defend the arguments in conflict.

Also, we can define restricted coalition admissible and restricted coalition semantics as we show above.

Definition 4.7 [Self-organizing coalition] A coalition C_i is a self-organizing coalition about argument $a \iff \forall$ agent $\alpha \in C_i$, $\forall$ argument $\beta \sqsubset \alpha$, a and β do not attack each other.

This definition shows which agents will tend to form a coalition to defend a specific argument.

In particular, we notice that: if coalition E defends a based on a self-organizing coalition about a, then E restricted coalition defends a.

5   Properties

Given a coalition argumentation framework $CAF = \langle \mathcal{A}, \to, \mathcal{S}, \sqsubset, C \rangle$, the coalition demonstrates different properties under different semantics as follows:

First, under stable semantics, since every argument that does not belong to an extension should be attacked by the extension, each possible coalition that makes non-empty extensions should be able to involve all arguments. There might be more than one coalition that will enforce an extension or a set of extensions. So, we may define a partial order over a set of coalitions in terms of some criteria, such as their size, authorities and values, etc.

Second, under grounded semantics, if $\langle \mathcal{A}, \to \rangle$ has no non-empty grounded extension, it is impossible to form a coalition that may enforce a non-empty grounded extension. Under preferred semantics, there exists at least one extension. So, there exists at least one coalition to enforce each preferred extension of $\langle \mathcal{A}, \to \rangle$. In this case, we may also define a partial order of coalitions.

Third, if we replace all the agents of a coalition by one agent, then we derive the same extensions under agent defense semantics.

6   Related work

Given a set of agents and a set of arguments where each agent may have several arguments and an argument may belong to several agents, we are interested in how agents have the ability to defend arguments in the form of coalitions. There are other variants of semantics that adapt these notions, such as weak defense for weak admissibility semantics [7], but that is not based on the agent metaphor. Arisaka and Satoh [3] adapt the notion of conflict-free to conflict-eliminability and then apply their four new coalition formability semantics to a Japanese political example, while in our paper, we use a concrete and realistic running example to show coalition defense and the corresponding new semantics. Kontarinis and Toni [19] analyse the identification of the malicious behavior of agents in the form of a bipolar argumentation framework which, together with the work of Panisson et al. [21], may inspire work of agent reduction semantics based on trustfulness. Our own recent work [24] involved a complete analysis of four types of semantics of agent argumentation.

There is a lot of early work on how to generate coalitions of agents with different mechanisms [1,13,18]. Boella et al. [8] developed social viewpoints for arguing about coalitions, considering attacks on attacks in the context of reasoning about coalitions. Amgoud [2] discusses task allocation via coalition formation. She points out that agents need to form a coalition to help each other in order to fulfill joint tasks better, and proposes a framework where several coalition structures can be generated and then evaluated by agents based on their preferences. She presents a proof theory that agents do not need to test the whole structure in order to check whether or not the given coalition is good. In our work, we pre-define the coalition of agents (only agents in the same coalition have the ability to defend each other's arguments) and then we evaluate the semantics of the framework through the new defense. Arisaka et al. [4] extend agent argumentation frameworks with coalitions among agents. Rienstra et al. [22] consider the case where agents may have different semantics, for example one agent uses grounded semantics and another agent uses preferred semantics. Bulling et al. [9] use Alternating-time temporal logic (ATL) as the technical basis for modelling coalition formation. They extend ATL for modelling coalitions through argumentation and reasoning about the abilities of coalitions of agents and the formation of coalitions, where coalition formation is part of the logical language. In our work, we model coalition formation in a different way by extending Dung's abstract argumentation, associating arguments with agents, and partitioning of the agents to represent coalitions. They define defense and conflict-free based on defeat rather than attack, and put forward a valid coalition in which members in a coalition are undefeated. In our work, the arguments proposed by agents in the same coalition can conflict.

Cayrol & Lagasquie-Schiex [11] propose a coalition of arguments instead of a coalition of agents in the setting of bipolar argumentation. They define a meta-framework consisting of a set of meta-arguments as well as their conflict relation such that an attack exists only at the meta-level. Support relation is used to relate members in a meta-argument. Their work differs from ours in how the coalition of arguments is conflict-free. Whereas they argue that arguments represent agents and only the ones who want to cooperate come together, in our work, we allow arguments belonging to agents who are in a coalition conflict. Another difference is that they follow Dung's methodology for defining semantics with the construction of a meta-argumentation framework, while our work defines coalition defense and coalition semantics.

To meet the maximum agreement, Leite & Martins [20] introduce an abstract model of argumentation where agents can vote for or against an issue. Another related work is the aggregation of individual views into collective acceptability, which in general is split into two directions: semantics aggregation and structured aggregation. Some authors [5,10] capture the notion that individual members need to defend collective decisions in order to reach a compatible outcome, and propose to address judgement aggregation by combining different individual evaluations (semantics) of the situation represented by an argumentation framework. On the other hand, Chen et al. [12] evaluate how to aggregate abstract argumentation frameworks with the preservation of semantics. Hunter et al. [16] take an epistemic approach to probabilistic argumentation where arguments are believed or not believed by varying degrees, providing an alternative to the subtle standard in Dung's framework. Due to space limitations, we focus on the form of coalitions, while we also inflect the maximum agreement through social semantics that prefers arguments that belong to more than one agent.

Another related work is concept accrual, *i.e.* arguments that cannot defeat their target on their own but can succeed together [23] can also increase the acceptability of their inner arguments. For example, let A, B, C be three arguments. A defeats B, A defeats C, neither B nor C can defeat A but B and C accrues to defeat A. Accrual differs from coalition. The former increases the acceptability of arguments through preferences or strength. It mainly concerns the defeat relation between arguments (accruals). The latter increases the acceptability of arguments through defense. It has nothing to do with defeat or preferences. In our future work, it may be valuable to incorporate accrual in CAF. The key points are:

- defining preferences among sets of arguments on the strength of each argument;
- prescribing that arguments can only accrue with arguments in the same coalition;
- identical arguments that belong to different agents can accrue to strengthen themselves.

In our work, we define coalitions based on agents, not arguments, for which we define the new notion of coalition defense. However, the semantics are the same. Our choice can more strongly indicate the meaning of a coalition in the sense used by society, i.e. agency.

7   Future work

The example can be considered in another way. In our example, the coalition helps the agents to jointly make their arguments stronger and increase the acceptability of the arguments they defend. Intuitively, the coalition may also make the arguments it defends weaker or decrease their acceptability.

The idea of persuasion in the work of Anthony Hunter [15], where he talks about persuasion using probability argumentation, may give us some insight. When one agent wants to persuade another, (s)he has uncertain information. We can use a coalition about a set of agents, and probably, we don't know with certainty which argument would be proposed by others, but we know the probability of that. We can do this by forming a coalition that has a higher probability of reaching some goals.

The main goal of collective argumentation is to achieve maximum agreement among a set of agents. Thus, another future work is to discuss how we can reach maximum agreement when we want to form a coalition.

In our paper, the running example is static, but presumably agents not only enter into coalitions but also leave them. This issue of dynamics can be addressed by modularity. When agents join or leave, we just consider the arguments that attack or are attacked by the arguments which belong to the changed agent. In this way, we can get new extensions.

We consider all the possibilities of making up coalitions. And one important thing is to evaluate different coalitions: which is better and what's the criterion? One potential way is to consider the stability of coalitions. If we think of agents taking a principled stand on issues, their positions dictate which coalitions they can enter into and sustain even as the coalitions expand. The strength of arguments can then be measured by the stability of coalitions coming together in support of those arguments and against counter-arguments. It is also related to the dynamics of coalition.

8   Conclusions

We have adapted Dung's notion of defense [14] to incorporate coalitions and proposed a coalition argumentation framework. In this formal framework, we mainly distinguished between two reasons for forming coalitions and find the properties of coalitions.

In this paper we investigated how coalition formation can be analyzed among agents that put forward arguments and try to persuade each other. Coalition formation is typically studied in a game theoretic setting, but we adopted the standard abstract model of argumentation introduced by Dung, we associated arguments with agents, and we introduced new coalition semantics.

We distinguished between three kinds of collaboration among arguments. First, agents may form coalitions to put forward the same arguments, if putting forward the same arguments increases the strength of such arguments in the debate. Second, agents may put forward arguments to defend the arguments of other agents in the coalition. Third, agents may work together to provide defenses for the attackers on their shared arguments.

In our formal approach, we focused on the second and third kinds of collaboration, because the first is already widely studied in social semantics using techniques from voting theory and social choice. We extended Dung's theory of abstract argumentation in two ways. First, we introduced agents and coalition structures in argumentation frameworks. Secondly, we adapted the notion of defense [14] such that arguments can only be defended by arguments from the same coalition. Everything else stayed exactly the same.

We showed how this theory can be used for coalition formation using an extended case study in political debate. With four agents and seven arguments, we showed how coalitions affect the accepted arguments for various kinds of coalition semantics. Considering the political debate example, if L doesn't collaborate with O, argument i will not be accepted.

Our investigation highlighted some properties that can be studied in a formal setting. We leave the formal analysis and the formalization of the formation of coalitions to the journal extension of this paper.

Acknowledgements

This work has received funding from the EU H2020 research and innovation programme under Marie Skłodowska-Curie Actions Innovative Training Networks European Joint Doctorate grant agreement No. 814177 in Law, Science and Technology, a Joint Doctorate on the Rights of Internet of Everything.

References

[1] Samir Aknine, Suzanne Pinson, and M Shakun. Coalition formation problem: New multi-agent methods with preference models. In *Proceedings of the AAAI workshop on Coalition formation in dynamic multi-agent environments*, pages 51–56, 2002.

[2] Leila Amgoud. An argumentation-based model for reasoning about coalition structures. In *International Workshop on Argumentation in Multi-Agent Systems*, pages 217–228. Springer, 2005.

[3] Ryuta Arisaka and Ken Satoh. Coalition formability semantics with conflict-eliminable sets of arguments. In *Proceedings of the 16th Conference on Autonomous Agents and MultiAgent Systems, AAMAS 2017, São Paulo, Brazil*, pages 1469–1471, 2017.

[4] Ryuta Arisaka, Ken Satoh, and Leendert W. N. van der Torre. Anything you say may be used against you in a court of law - abstract agent argumentation (triple-a). volume 10791, pages 427–442. Springer, 2017.

[5] Edmond Awad, Richard Booth, Fernando Tohmé, and Iyad Rahwan. Judgement aggregation in multi-agent argumentation. *J. Log. Comput.*, 27(1):227–259, 2017.

[6] Haris Aziz, Felix Brandt, Edith Elkind, and Piotr Skowron. Computational social choice: The first ten years and beyond. In *Computing and software science*, pages 48–65. Springer, 2019.

[7] Ringo Baumann, Gerhard Brewka, and Markus Ulbricht. Comparing weak admissibility semantics to their Dung-style counterparts–reduct, modularization, and strong equivalence in abstract argumentation. In *Proceedings of the International Conference on Principles of Knowledge Representation and Reasoning*, volume 17, pages 79–88, 2020.

[8] Guido Boella, Leendert Van Der Torre, and Serena Villata. Social viewpoints for arguing about coalitions. In *Pacific Rim International Conference on Multi-Agents*, pages 66–77. Springer, 2008.

[9] Nils Bulling, Jürgen Dix, and Carlos Iván Chesnevar. Modelling coalitions: Atl+ argumentation. In *AAMAS (2)*, pages 681–688, 2008.

[10] Martin Caminada and Gabriella Pigozzi. On judgment aggregation in abstract argumentation. *Autonomous Agents and Multi-Agent Systems*, 22(1):64–102, 2011.

[11] Claudette Cayrol and Marie-Christine Lagasquie-Schiex. Coalitions of arguments: A tool for handling bipolar argumentation frameworks. *Int. J. Intell. Syst.*, 25(1):83–109, 2010.

[12] Weiwei Chen and Ulle Endriss. Preservation of semantic properties in collective argumentation: The case of aggregating abstract argumentation frameworks. *Artificial Intelligence*, 269:27–48, 2019.

[13] Viet Dung Dang and Nicholas R. Jennings. Generating coalition structures with finite bound from the optimal guarantees. In *3rd International Joint Conference on Autonomous Agents and Multiagent Systems (AAMAS 2004)*, pages 564–571. IEEE Computer Society, 2004.

[14] Phan Minh Dung. On the acceptability of arguments and its fundamental role in nonmonotonic reasoning, logic programming and n-person games. *Artificial intelligence*, 77(2):321–357, 1995.

[15] Anthony Hunter. Modelling the persuadee in asymmetric argumentation dialogues for persuasion. In Qiang Yang and Michael J. Wooldridge, editors, *Proceedings of the Twenty-Fourth International Joint Conference on Artificial Intelligence, IJCAI 2015, Buenos Aires, Argentina, July 25-31, 2015*, pages 3055–3061. AAAI Press, 2015.

[16] Anthony Hunter, Sylwia Polberg, and Matthias Thimm. Epistemic graphs for representing and reasoning with positive and negative influences of arguments. *Artif. Intell.*, 281:103236, 2020.

[17] Nikos Karanikolas, Pierre Bisquert, and Christos Kaklamanis. A voting argumentation framework: Considering the reasoning behind preferences. In *ICAART 2019-11th International Conference on Agents and Artificial Intelligence*, volume 1, pages 42–53. SCITEPRESS-Science and Technology Publications, 2019.

[18] Matthias Klusch and Andreas Gerber. Dynamic coalition formation among rational agents. *IEEE Intelligent Systems*, 17(3):42–47, 2002.

[19] Dionysios Kontarinis and Francesca Toni. Identifying malicious behavior in multi-party bipolar argumentation debates. In *Multi-Agent Systems and Agreement Technologies - 13th European Conference*, volume 9571 of *Lecture Notes in Computer Science*, pages 267–278. Springer, 2015.

[20] João Leite and João G. Martins. Social abstract argumentation. In Toby Walsh, editor, *IJCAI 2011, Proceedings of the 22nd International Joint Conference on Artificial Intelligence, Barcelona, Catalonia, Spain, July 16-22, 2011*, pages 2287–2292. IJCAI/AAAI, 2011.

[21] Alison R. Panisson, Simon Parsons, Peter McBurney, and Rafael H. Bordini. Choosing appropriate arguments from trustworthy sources. In *Computational Models of Argument - Proceedings of COMMA 2018, Warsaw, Poland, 12-14 September 2018*, volume 305, pages 345–352. IOS Press, 2018.

[22] Tjitze Rienstra, Alan Perotti, Serena Villata, Dov M Gabbay, and Leendert van der Torre. Multi-sorted argumentation. In *International Workshop on Theorie and Applications of Formal Argumentation*, pages 215–231. Springer, 2011.

[23] Julien Rossit, Jean-Guy Mailly, Yannis Dimopoulos, and Pavlos Moraitis. United we stand: Accruals in strength-based argumentation. *Argument & Computation*, 12:87–133, 2020.

[24] Liuwen Yu, Dongheng Chen, Lisha Qiao, Yiqi Shen, and Leendert van der Torre. A principle-based analysis of abstract agent argumentation semantics. 2021. under review.

A Logical Description of Strategizing in Social Network Games

Ramit Das [1]

Institute of Mathematical Sciences and
Homi Bhabha National Institute
Chennai, India

R Ramanujam [2]

Institute of Mathematical Sciences and
Homi Bhabha National Institute
Chennai, India

Abstract

We propose a modal logic for reasoning about strategies in social network games ([12]). In these games, players are connected by a social network graph, and payoffs for players are determined by choices of players in their neighbourhood. We consider improvement dynamics of such games and the formulas of the logic are intended to capture bisimulation classes of improvement graphs. The logic is structured in two layers: local formulas which specify neighbourhood dependent strategization, and global formulas which describe improvement edges and paths. Notions like Nash equilibrium and (weak) finite improvement property are easily defined in the logic. We show that the logic is decidable and that the valid formulas admit a complete axiomatization.

Keywords: Logic of strategies, Social network games, Threshold reasoning, Graphical games, Decidability

1 Background

There has been extensive work on the logical foundations of game theory in the last couple of decades. [13] presents an excellent survey of the logical issues in reasoning about games. Asserting the existence of equilibria and exploring the underlying rationality assumptions forms the crux of many of these studies ([6], [16], [1]). On the other hand, much of game theory studies the *existence* of strategies and the logical approach has led to studying compositional structure in strategies ([10], [15], [2]).

[1] ramitd@imsc.res.in

[2] jam@imsc.res.in

A fundamental assumption of non-cooperative game theory is that players strategize individually and independent of each other. This was referred to as the *Great Simplification* by von Neumann in 1928 and indeed, this is what leads to the abstraction of normal form games. However, such a flat structure on the set of players is not always realistic. In the case of social networks like Facebook and Twitter, individuals are influenced by their friends, and often seek to influence their friends, in the choices they make. The 'payoff' in such interactive behaviour is often social, in the sense that matching one's friends' choices may indeed be the desired outcome. Such *majority games* (and their dual, so-called *minority games*) are also extensively discussed in the literature. 'Facebook logic' ([11]) and its counterparts discuss such relationships and their impact on decision making.

A specific class of such games on social networks was studied by Apt and Simon ([12]), which is the starting point of departure for our paper. While they study the complexity of computing equilibria in such games, we take their *improvement dynamics* and seek a logical description. The central question we take up is this: how do we abstract away from the details of utilities and preferences, and get to the core of strategization by players in such games?

A natural way for such abstraction is to consider game equivalences and seek logical descriptions of equivalence classes. When outcomes are determined locally, by neighbourhoods in the social network, this induces further structure in the improvement graphs, which leads to interesting bisimulation classes. This naturally leads us to a modal logic as a tool for strategization structure.

The logic we present has two crucial points of departure from other modal descriptions of games: it describes improvement dynamics rather than player preferences; and it describes *threshold reasoning* in strategic choice. The latter is only intended as an instance of local strategization, and other similar logical means would be equally suitable.

We go on to present a complete axiomatization of the logic and show that the satisfiability problem is decidable. These technical results suggest that we have an interesting logical formalism at hand. However, only by specifying different classes of network games and reasoning about their improvement dynamics, can we develop this theory further.

2 Social network games

Fix $[n] = \{1, \ldots, n\}$ a fixed, finite set of players (or agents), and we talk of n-player games. Let Σ denote a set of *choices* or *strategies* available to a player. We only consider games in which all players have the same set of choices. This is for convenience of presentation. [3] We let i, j, etc index players and a, b, etc range over Σ. A *strategy profile* is an element of Σ^n, and we let σ, σ' etc to range over strategy profiles with $\sigma[i]$ denoting the ith element of σ, which is interpreted as the choice of player i.

---

[3] The material here could be developed with specific choice sets for each player, but this clutters up the formalism without adding significant insight.

Let Ω be a set of *outcomes* (or *payoffs*) and a utility function (or *payoff* function) is a map: $\pi : \Sigma^n \to \Omega^n$. It is assumed that Ω is a partially ordered set, with the order $\preceq$.[4]

Definition 2.1 A game is a tuple $H = (n, \Sigma, \Omega, \pi)$, fixing the number of players, the strategy sets of players, the outcomes and the payoff map.

These are called strategic form (or normal form) games, extensively studied for nearly a century. In games on social networks we assume a graph structure on the set of players, and the edge relation is interpreted to be a form of friendship relation that governs behaviour in a suitable way ([12]).One reason for studying such *graphical games* is that the payoff function above is large, being exponential in n and hence hard to present. In social networks, though the number of players may be large, each player interacts with only a few players. It is often possible to assume that each player, on average, interacts with at most *log n* many players where n is the number of players in the game. In this case, we can consider games where the payoff is determined only by player *neighbourhoods* and the payoff function is then, roughly, of size $\Sigma^{log\ n}$.

Definition 2.2 A **social network game** is a tuple $G = (n, \Sigma, \Omega, E, \pi)$, where $E \subseteq ([n] \times [n])$ is the edge relation of the social network graph, $\pi = (\pi_1, \ldots, \pi_n)$ is the payoff function, one for each player, where $\pi_i : \Sigma^{|N_i|} \to \Omega$, $N_i = \{i\} \cup \{j \mid (j, i) \in E\}$, is the *neighbourhood* of player i.

Clearly π induces a function from $\Sigma^n \to \Omega^n$ which, by abuse of notation, we again denote by π.

Typically, a social network graph is assumed to be simple and undirected: that is, the edge relation is symmetric and has no self-loops. When the edge relation represents a form of friendship, it is surely bi-directional. But we prefer to retain the more general form of directed graphs. For instance the edge from i to j may represent i linking to j on the web, in which case, there is no reason to assume a link in the reverse direction. Self-loops do not matter since we have included every player in its neighbourhood by definition.

How does a player *strategize* in such a game? This clearly depends on the connectivity in the social network graph. For instance, when $E = [n] \times [n]$ it is just the same as reasoning in normal form games.

Consider the edge relation $E_1 = \{(1, 2), (1, 3), (2, 3)\}$. Here $N_1 = \{1\}$, $N_2 = \{1, 2\}$ and $N_3 = \{1, 2, 3\}$. In this game, player 1 makes choices independent of others since her payoff depends only on what she chooses. Player 2 provides his best response to 1's choices, and player 3 provides a best response to every combination of choices of players 1 and 2.

2.1 Modelling examples

Threshold based reasoning is common in games on social networks ([8]), and we present two examples of such modelling.

---

[4] Again, for convenience, we assume a uniform ordering on outcomes rather than one order $\preceq_i$ for each player i.

The first is that of a **Majority Game**, used in modelling social phenomena such as voting: ([3], [7]). $\Sigma = \{0, 1\}$. Let $\sigma = (a_1, \dots, a_n)$ be any profile. The payoff for player i at σ is 1 if $\frac{|\{j \in N(i) | a_j = a_i, j \neq i\}|}{|N(i)|} > \frac{1}{2}$, and is 0 otherwise. The trivial equilibria for this game are all players choosing 0, or all choosing 1.

A non-trivial vote will be the following:

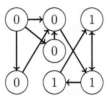

Fig. 1. Non trivial Nash Equilibrium in the Majority Game

For another example, consider the **"Best Shot" Public Goods game** ([8]). Again, $\Sigma = \{0, 1\}$. In this game, there is an option of taking an altruistic action for the public good, or refraining from it. Doing good carries a fixed uniform cost $c \in (0, 1)$. Of course, if some neighbour takes action, it is much better and one can enjoy the result doing nothing. Alas, if everyone thinks so, nobody benefits. This is specified by the payoff function as follows. Again let $\sigma = (a_1, \dots, a_n)$:

$$\pi_i(\sigma) = \begin{cases} 0 & \text{if } a_i = 0, a_j = 0 \text{ for all } j \in N(i) \\ 1 & \text{if } a_i = 0, a_j = 1 \text{ for some } j \in N(i) \\ 1 - c & \text{if } a_i = 1 \end{cases}$$

Here is a Nash equilibrium for this game:

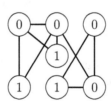

Fig. 2. Nash Equilibrium in the Public Goods Game

2.2 Dynamics

A simple way to study such reasoning is given by the improvement graph dynamics defined as follows.

Definition 2.3 The **improvement graph** I_G, associated with the game G, is the graph $I_G = (\Sigma^n, IE_G)$, where $IE_G \subseteq (\Sigma^n \times [n] \times \Sigma^n)$ is the player labelled edge relation given by $(\sigma, i, \sigma') \in IE_G$ iff $\pi(\sigma)[i] \prec \pi(\sigma')[i]$ and for all $j \neq i$, $\sigma[j] = \sigma'[j]$.

We have an i-labelled edge from a strategy profile to another, if player i can unilaterally deviate from the former to the latter to get an improved payoff.

Note that at a profile, there can be different i-improvement edges leading to different profiles (with perhaps incomparable outcomes). A path in I_G is an improvement path.

Note that the improvement graph can have cycles. For instance, consider the two-player game of matching pennies: both players call heads or tails, the first player wins when the results match, and the second wins when there is a mismatch. We then have the cycle $(H, H) \to_2 (H, T) \to_1 (T, T) \to_2 (T, H) \to_1 (H, H)$.

In any particular game n is fixed as well as the size of Σ and hence I_G is a finite directed graph, though a large one (its size being exponential in n). It contains a good deal of interesting information about the game G. For instance consider a sink node of I_G, which has no out-going edge: it is easy to see that a strategy profile is a sink node if and only if it constitutes a *Nash equilibrium*, from which no player has any incentive to deviate.

I_G includes structural information from the social network graph as well. For instance, we have the following proposition.

Proposition 2.4 *Let G be a social network game and players i, j such that $N_i \cap N_j = \emptyset$. Then, for all $\sigma_0, \sigma_1, \sigma_2 \in \Sigma^n$ we have:*

- *If $\sigma_0 \to_i \sigma_1$ and $\sigma_0 \to_j \sigma_2$, then there exists $\sigma_3 \in \Sigma^n$ such that $\sigma_1 \to_j \sigma_3$ and $\sigma_2 \to_i \sigma_3$.*

- *If $\sigma_0 \to_i \sigma_1$ and $\sigma_1 \to_j \sigma_2$, then there exists $\sigma_3 \in \Sigma^n$ such that $\sigma_0 \to_j \sigma_3$ and $\sigma_3 \to_i \sigma_2$.*

To see this, fix G as above and consider profiles $\sigma_0, \sigma_1, \sigma_2 \in \Sigma^n$ such that $\sigma_0 \to_i \sigma_1$ and $\sigma_0 \to_j \sigma_2$. Firstly note that for all $k \in [n]$, $k \neq i$ and $k \neq j$, $\sigma_1[k] = \sigma_2[k]$. Define σ_3 by: $\sigma_3[i] = \sigma_1[i]$, $\sigma_3[j] = \sigma_2[j]$, and for all $k \in [n]$, $k \neq i$ and $k \neq j$, $\sigma_3[k] = \sigma_0[k]$. Note that $\pi(\sigma_0)[i] = \pi(\sigma_2)[i] \prec \pi(\sigma_3)[i]$, and $\pi(\sigma_0)[j] = \pi(\sigma_1)[j] \prec \pi(\sigma_3)[j]$, as required since $N_i \cap N_j = \emptyset$ and π_k is dependent only on N_k, for all players k. The other statement in the proposition is proved similarly.

The proposition refers to 'squares' in the improvement graph. In general if we have k players with pairwise disjoint sets of neighbourhoods, we have k-hypercubes embedded in I_G. This may be interpreted as concurrent strategization by players in the game. Such structure has been extensively studied in asynchronous transition systems in concurrency theory ([9]).

In the analysis of games, we are typically interested in questions like whether the game has a Nash equilibrium, whether every improvement path is finite, whether an equilibrium profile is reachable from every profile, etc. Algorithmically all these questions are efficiently solvable, but then the input, namely the improvement graph, is itself large.

Note that the improvement graph dynamics induces a natural **game equivalence**: we can consider two games to be equivalent if they have a similar (but not necessarily isomorphic) improvement structure.

Definition 2.5 *Let G, G' be two n player games with strategy sets Σ and Σ',* a relation $R \subseteq (\Sigma^n, \Sigma'^n)$ is an improvement bisimulation if whenever $(\sigma, \sigma') \in R$, for all $i \in [n]$,

- whenever $\sigma \to_i \sigma_1$ in I_G, there exists profile σ'_1 in game G' such that $\sigma' \to_i \sigma'_1$ in $I_{G'}$ and $(\sigma_1, \sigma'_1) \in R$.
- whenever $\sigma' \to_i \sigma'_1$ in $I_{G'}$, there exists profile σ_1 in game G such that $\sigma \to_i \sigma_1$ in I_G and $(\sigma_1, \sigma'_1) \in R$.

The relation is on games in general rather than on social network games. Clearly, the improvement bisimulation relation is an equivalence relation. We say that G and G' are bisimilar if there exists a nonempty bisimulation on their improvement graphs. When we reduce strategic form games by this equivalence, we abstract from specific outcomes as well as player strategies but preserve the *player strategisation structure*. Note that outcomes and orderings on them have entirely disappeared in the bisimulation classes, only the information that some improvement in outcome is possible (or not) is retained. This is a semantic characterization, and we need structural constraints to capture the semantic conditions. What we would like to do is to study player rationale to provide logical structure to the strategization. In the context of social network games, we use threshold reasoning over player neighbourhoods as a way of specifying this rationale.

3   The logic

When we seek a logical description of improvement dynamics, it is natural to consider first order logics and their extensions with least fixed-point operators, since equilibrium computation typically proceeds by finding such fixed points. In earlier work ([4]) we have carried out such an exercise. However, in light of the discussion above, we are not interested in the improvement graphs themselves but in bisimulation classes, and hence define a modal logic for this study. Player strategisation involves reasoning about strategies played by other players in their neighbourhood. Thus mutual strategization by players becomes relevant, and the logic we define below includes precisely such *local* reasoning by players as well as *global* improvement dynamics.

Syntax The formulas of the logic are presented in two layers: **local player formulas** and **global outcome formulas**. The logic is parameterised by n, the number of players, and the strategy set Σ.

The syntax of local formulas is given by:

$$\alpha \in L_i ::= a \in \Sigma \mid e_j \mid \neg \alpha \mid \alpha \vee \alpha' \mid N_{rel\ r}\ \alpha$$

where $rel \in \{\geq, \leq, <, >\}$, $i \in [n]$ and r is a rational number, $0 \leq r \leq 1$.

The atomic formula a asserts that player i chooses a, and the atomic formula e_j asserts that there is a directed edge from j to i, that player i is dependent on j. $N_{rel\ r}\alpha$ considers the size of the neighbourhood choosing α: for instance, $N_{\leq r}\alpha$ asserts that at most an r-fraction of players in the neighbourhood of i support α.

Fix P a finite set of atomic propositions denoting *conditions on outcomes*. These are qualitative outcomes, used to denote levels of satisfaction. We will characterize outcomes by sets of propositions, which can be equivalently thought of as boolean formulas on P.

The syntax of global formulas is given by:

$$\phi \in \Phi ::= p@i,\ p \in P \mid \alpha@i, \alpha \in L_i \mid \neg\phi \mid \phi \vee \phi' \mid \langle i\rangle\phi \mid \Diamond^*\phi$$

The global formulas constitute a standard propositional modal logic of transitive closure built over local formulas. Note that the atomic formulas $p@i$ and $\alpha@i$ are of different sort: the former refers to outcomes and the latter to strategies. The other boolean connectives $\wedge$, $\supset$ and $\equiv$, for conjunction, implication and equivalence are defined in the standard manner, for both local and global formulas. The dual formulas are: $[i]\phi = \neg\langle i\rangle\neg\phi$ and $\Box^*\alpha = \neg\Diamond^*\neg\phi$ We use the abbreviation $\bigcirc\phi = \bigvee_{i\in[n]} \langle i\rangle\phi$ and $\odot\phi = \neg\bigcirc\neg\phi$. (We use $\top$ and $\bot$ to refer to the propositional constants 'True' and 'False' which are coded by $p@i \vee \neg p@i$ and $p@i \wedge \neg p@i$, for a fixed propositional symbol p.)

Semantics The formulas are interpreted over strategy profiles of social network games. A model is a social network game $M = (n, \Sigma, 2^P, E, \pi)$ where $\Omega = 2^P$ is the set of outcomes. The ordering can be seen as an ordering on boolean formulas on P.

The semantics is given by assertions of the form $M, \sigma \models \phi$, read as ϕ is true of the strategy profile σ in model M. This in turn depends on the satisfaction relation for local formulas. For $i \in [n]$ and $\alpha \in L_i$, we define i-local satisfaction relations:

- $M, \sigma \models_i a$ if $\sigma[i] = a$.
- $M, \sigma \models_i e_j$ if $(j, i) \in E$.
- $M, \sigma \models_i \neg\alpha$ if $M, \sigma \not\models_i \alpha$.
- $M, \sigma \models_i \alpha \vee \beta$ if $M, \sigma \models_i \alpha$ or $M, \sigma \models_i \beta$.
- $M, \sigma \models_i N_{rel\ r}\ \alpha$ if $\frac{|\{j\mid M,\sigma\models_j\alpha\}|}{|N_i|}$ rel r.

The semantics of global formulas can then be defined as follows. Below, let $\rightarrow^* = (\cup_i \rightarrow_i)^*$, the reflexive transitive closure of the union of the improvement edge relations.

- $M, \sigma \models p@i$ if $p \in \pi_i(\sigma)$.
- $M, \sigma \models \alpha@i$ if $M, \sigma \models_i \alpha$.
- $M, \sigma \models \neg\phi$ if $M, \sigma \not\models \phi$.
- $M, \sigma \models \phi \vee \psi$ if $M, \sigma \models \phi$ or $M, \sigma \models \psi$.
- $M, \sigma \models \langle i\rangle\phi$ if there exists σ' such that $\sigma \rightarrow_i \sigma'$ and $M, \sigma' \models \phi$.
- $M, \sigma \models \Diamond^*\phi$ if there exists σ' such that $\sigma \rightarrow^* \sigma'$ and $M, \sigma' \models \phi$.

We say that ϕ is satisfiable if there exists a social network game model

M and a profile σ such that $M, \sigma \models \phi$. We say that ϕ is valid if $\neg\phi$ is not satisfiable.

It is easy to see that Nash Equilibrium is given by the simple formula:
$NE = \bigwedge_{i \in [n]} [i]\perp$. To assert that there is a path from the current profile to a Nash
Equilibrium, we write: $\Diamond^*(NE)$. To assert the "Weak Finite Improvement Property", that a Nash Equilibrium profile is reachable from every profile, we write: $\Box^*\Diamond^*(NE)$.

The strategy specification for the majority game is simple. Let the payoff set be given by: $P = \{p_0, p_1\}$ with $\{p_0\} \preceq \{p_1\}$. The formula $(N_{>\frac{1}{2}}(e_j \wedge 1))@i \equiv p_1@i$ defines the payoff map.

For the Public Goods Game, let $P = \{p_0, p_c, p_1\}$ with $\{p_0\} \preceq \{p_c\} \preceq \{p_1\}$. The payoff map is specified by:

$$(\bigwedge_i 0@i) \supset (\bigwedge_i p_0@i) \wedge \bigwedge_i (1@i \supset p_c@i) \wedge \bigwedge_i ((0 \wedge N_{>0}1)@i \supset p_1@i)$$

4   Axiomatization and decidability

We now present an axiomatization of the valid formulas. We have one axiom system Ax_i for each player i in the system, and in addition a global axiom system AX to reason about improvement dynamics. In some sense, this helps to isolate how much global reasoning is required.

Below, we say rel r *entails* rel' r' when $r \leq r'$ and either $rel = rel' =\leq$ or $rel =<$ and $rel' =\leq$, or $r \geq r'$ and either $rel = rel' =\geq$ or $rel =>$ and $rel' =\geq$. Further we say rel' is the *complement* of rel if one of them is $\geq$ and the other is $<$, or one is $\leq$ and the other $>$.

We use the notation $\vdash_i \alpha$ to mean that the formula $\alpha \in L_i$ is a theorem of system Ax_i. Similarly, $\vdash \phi$ means that ϕ is a theorem of the global system.

Ax_i, The axiom schemes for agent i

($A0_i$) All the substitutional instances of propositional tautologies
($A1_i$) $N_{rel\ r}(\alpha \supset \beta) \supset (N_{rel\ r}\alpha \supset N_{rel\ r}\beta)$
($A2_i$) $\alpha \supset N_{>0}\alpha$
($A3_i$) $N_{rel\ r}\alpha \equiv \neg N_{rel'\ r}\alpha$, rel' complement rel
($A4_i$) $N_{rel\ r}\alpha \supset N_{rel'r'}\alpha$, rel r entails rel' r'

Inference rules

$(MP_i)\ \dfrac{\alpha,\ \ \alpha \supset \beta}{\beta}\qquad (NG_i)\ \dfrac{\alpha}{N_{\geq 1}\alpha}$

The axioms of the local system are quite standard. The Kripke axiom applies to every instance of the $N_{rel\ r}$ modality, and the remaining axioms express properties of inequalities. The rule (NG_i) reflects the fact that properties which are invariant in the system hold throughout the neighbourhood.

In the global axiom system. we have Kripke axioms for $[i]$ modalities and for transitive closure, and an induction rule. We need a "transfer" rule to infer $\alpha@i$ globally when we infer α locally. The remaining axioms relate to social network games: specifying the fact that formulas are asserted at strategy profiles, corresponding to one choice for each player, that payoffs for a player are determined by the player's neighbourhood, and so on.

Global axiom schemes AX

$(B0)$ $(\neg\alpha)@i \equiv \neg\alpha@i$

$(B1)$ $(\alpha \vee \beta)@i \equiv (\alpha@i \vee \beta@i)$

$(B2)$ $[i](\phi_1 \supset \phi_2) \supset ([i]\phi_1 \supset [i]\phi_2)$

$(B3)$ $\Diamond^*\phi \equiv (\phi \vee \bigcirc\Diamond^*\phi)$

$(B4)$ $\alpha@j \equiv [i]\alpha@j, \quad j \neq i$

$(B5)$ $(p@j \equiv [i]p@j) \wedge (\neg p@j \equiv [i]\neg p@j) \quad i \notin N_j$

$(B6)$ $(\bigwedge_{j \in N_i} a_j@j) \supset$

$$((p@i \supset \Box^*(\bigwedge_{j \in N_i} a_j@j \supset p@i)) \wedge (\neg p@i \supset \Box^*(\bigwedge_{j \in N_i} a_j@j \supset \neg p@i))$$

$(B7)$ $e_j@i \equiv \bigcirc e_j@i$

$(B8)$ $\Box^* \bigwedge_{i \in [n]} (\bigvee_{a \in \Sigma} (a@i \wedge \bigwedge_{b \neq a} \neg b@i))$

$(B9)$ $(N_{rel}\ r\alpha)@i \supset \bigvee_{J,K \subseteq N_i} (\bigwedge_{j \in J} \alpha@j \wedge \bigwedge_{k \in K} \neg\alpha@j) \quad K = N_i - J, \dfrac{|J|}{|J \cup K|}\ rel\ r$

Inference rules

(MP) $\dfrac{\phi, \quad \phi \supset \psi}{\psi}$ (GG) $\dfrac{\vdash_i \alpha}{\alpha@i}$ (G_i) $\dfrac{\phi}{[i]\phi}$

$(Conc)$ $\dfrac{\gamma_1 \vee \ldots \vee \gamma_\ell \quad (N_i \cap N_j = \emptyset)}{[\langle i\rangle\phi \wedge \langle j\rangle\psi] \supset \bigvee_{1 \leq k \leq \ell} [\langle i\rangle(\phi \wedge \langle j\rangle\gamma_k) \wedge \langle j\rangle(\psi \wedge \langle i\rangle\gamma_k)]}$

(Ind) $\dfrac{\psi \supset (\phi \wedge \bigcirc\psi)}{\psi \supset \Box^*\phi}$

The global axioms (B4) and (B5) assert that an i-improvement does not affect the strategies of other players, and hence the payoffs to players that do not have i in their neighbourhoods are unaffected. (B6) asserts that the payoff for a player i is determined only by the strategies of players in the neighbourhood of i. (B7) is a sanity check, that the social network graph is unaltered by improvement dynamics. (B8) asserts that the formulas are asserted over strategy profiles, with every player making a definite choice. (B9) asserts the correctness of neighbourhood threshold formulas.

The rule (Conc) asserts that players can concurrently improve if their neighbourhoods are disjoint, asserting the existence of a square in the improvement graph. This rule typifies the pattern of reasoning in a "true concurrency" based logic.

Proposition 4.1 *Every theorem of AX is valid.*

The soundness of the axioms mostly follow by the semantic definitions. (B4) follows from the fact that when $\sigma \rightarrow_i \sigma'$, $\sigma[j] = \sigma'[j]$. (B5) and (B6) follow from the definition of π_i. (B7) is valid since the social network does not vary with profiles. (B8) follows from the definition of strategy profiles. (B9) follows from the semantics of $N_{rel\ r}$ modality.

Among the rules, only the soundness of rule (Conc) is interesting. Assume the validity of the disjunction in the premise, and let $M, \sigma \models \langle i \rangle \phi \wedge \langle j \rangle \psi$. Let $\sigma \rightarrow_i \sigma_1$ and $\sigma \rightarrow_j \sigma_2$. By Proposition 2.4, there exists a profile σ_3 such that $\sigma_1 \rightarrow_j \sigma_3$ and $\sigma_2 \rightarrow_i \sigma_3$. Clearly, for some k, $M, \sigma_3 \models \gamma_k$. Hence $M, \sigma_1 \models \phi \wedge \langle j \rangle \gamma_k$, and $M, \sigma_2 \models \psi \wedge \langle i \rangle \gamma_k$. Thus, $M, \sigma \models \langle i \rangle (\phi \wedge \langle j \rangle \gamma_k) \wedge \langle j \rangle (\psi \wedge \langle i \rangle \gamma_k)$, as required.

Theorem 4.2 *AX provides a complete axiomatization of the valid formulas. Satisfiability of a formula ϕ can be decided in nondeterministic exponential time $(2^{O(m \cdot n)}$, where m is the length of ϕ and n is the number of players).*

Proof.

Call a formula ϕ *consistent* if $\not\vdash \neg\phi$. Call $\alpha \in L_i$ *i-consistent* if $\not\vdash_i \neg\alpha$. A finite set of formulas A is consistent if the conjunction of all formulas in A, denoted $\hat{A}$, is consistent. When we have a finite family S of sets of formulas, we write $\tilde{S}$ to denote the disjunction of all formulas $\hat{A}$, where $A \in S$.

For completeness, it suffices to prove that every consistent formula is satisfiable. In fact we show that every consistent formula ϕ is satisfiable in a model of size $2^{O(m \cdot n)}$ where m is the length of ϕ and n is the number of players. From this and soundness, we see a bounded model property: that every satisfiable formula is satisfiable in a model of size exponentially bounded in the size of the formula. This property at once gives a nondeterministic exponential time decision procedure for the logic as well.

Fix a given consistent formula ϕ_0. We confine our attention only to the subformulas of ϕ_0, and maximal consistent sets of subformulas. Towards this, for any i-local formula $\alpha \in L_i$, let $SF_i(\alpha)$ denote the set of subformulas of α. We assume it to be negation closed and to contain α. $|SF_i(\alpha)| = O(|\alpha|)$. Similarly, for any global formula ϕ, define $SF(\phi)$ to be the set of subformulas of ϕ, which is again negation closed and contains ϕ; further, if $\Diamond^*\psi \in SF(\phi)$ then so also $\bigcirc\Diamond^*\psi \in SF(\phi)$. Again, $|SF(\phi)| = O(|\phi|)$.

Let $R \subseteq SF(\phi_0)$. We call R an **atom** if it is a maximal consistent subset (MCS) of $SF(\phi_0)$. Note that, by rule (GG), for any atom R, $A_i = \{\alpha \mid \alpha@i \in R\}$ is i-consistent. Let $(A_1, \ldots, A_n)$ be the tuple of 'local atoms' in R. Let AT denote the set of all atoms. Define $\rightarrow \subseteq (AT \times [n] \times AT)$ by: $R_1 \rightarrow_i R_2$ iff $\{\phi \mid [i]\phi \in R_1\} \subseteq R_2$. Note that when $R_1 \rightarrow_i R_2$ and $\Box^*\phi \in R_1$, $\{\phi, \Box^*\phi\} \subseteq R_2$. Let $G_0 = (AT, \rightarrow)$.

Since ϕ_0 is consistent, there exists an MCS $R_0 \in AT$ such that $\phi_0 \in R_0$. Let G_1 be the induced subgraph of G_0 by restricting to atoms reachable from A_0, denoted $(AT_1, \rightarrow)$. We have the following observations on G_1.

- Every R in AT_1 induces a profile σ_R over $[n]$.
- For every R, R' in AT_1, and $i \in [n]$, if $\{a_j@j \in R \mid j \in N_i\} = \{a_j@j \in R' \mid$

$j \in N_i\}$ then $p@i \in R$ iff $p@i \in R'$.

- For every R in AT_1, if A_i is the i-local atom of R, and $N_{rel \ r}\alpha \in A_i$, then $|\{j \mid e_j \in A_i, \alpha \in A_j\}| \ rel \ r \cdot |N_i|$.

- For every R in AT_1, if $\langle i \rangle \phi \in R$ then there exists $R' \in AT_1$ such that $R \rightarrow_i R'$, $\phi \in R'$ and the boolean outcome formula in R' is higher in the preference ordering than the one in R.

- For every R in AT_1, if $\diamond^* \phi \in R$, then there exists an atom R' in AT_1 reachable from R such that $\phi \in R'$.

Axioms (B8) and (B6) ensure the first two conditions, the third uses the local axiom systems. The (Conc) rule ensures the fourth condition that when $\langle i \rangle \psi \in R$, we can indeed "compute" the maximal consistent set R' such that $R \rightarrow_i R'$. The last condition requires an argument such as the one used for propositional dynamic logic.

Define the game $G_{\phi_0} = (n, \Sigma, \Omega, E, \pi)$ by: $\Omega = 2^{P_0}$ where $P_0 = \{p \mid p@i \in SF(\phi_0), i \in [n]\}$; $E = \{(j, i) \mid e_j \in A_i \text{ of } R_0\}$; $\pi_i(a_{j_1}, \ldots, a_{j_k}) = \{p@i \in R \in AT_1 \mid a_{j_1}@j_1, \ldots, a_{j_k}@j_k \in R\}$, where $k = |N_i|$ and $\{\top@i\}$ if no such atom R exists, where $\top$ stands for "True" Note that for every profile that occurs in G_1, the payoff map is non-trivial. It is well-defined, by the second condition above. We have a model $M_{\phi_0} = (n, \Sigma, 2^{P_0}, E, \pi)$.

Then we show that for every subformula ϕ and every global maximal consistent set $R \in AT_1$, $\phi \in R$ iff $M_{\phi_0}, \sigma_R \models \phi$. This is proved by induction on the structure of ϕ. The axiom system ensures that neighbourhood specifications are consistent across players and the (Conc) rule ensures that when $\langle i \rangle \psi \in R$, we can indeed "compute" the maximal consistent set R' such that $R \rightarrow_i R'$.

Since there exists a maximal consistent set $R_0 \in AT_1$ such that the given formula $\phi_0 \in A_0$, we now have: $M_0, \sigma_{R_0} \models \phi_0$ and we are done.

$\square$

Thus we have completeness of axiomatization as well as decidability of satisfiability. While we have presented a non-deterministic exponential time decision procedure, we believe that it can be improved to deterministic exponential time: the main idea is to construct the entire atom graph but avoid guessing a good subgraph, but instead delete nodes and edges until what remains is a good subgraph.

Theorem 4.3 *The satisfiability problem is DEXPTIME-hard.*

This is proved along the lines of the lower bound for propositional dynamic logic, using the *finite corridor tiling problem*.

The model checking problem asks, given a social network game $M = (n, \Sigma, 2^P, E, \pi)$, and a formula ϕ whether there is some profile σ such that $M, \sigma \models \phi$. This can be solved in time $2^{O(n)} \cdot |E| \cdot |\phi|$, by explicitly constructing the strategy space and then running a standard labelling algorithm. This is exponential in the number of players, which is unavoidable since computing Nash equilibrium in social network games is known to be NP-hard ([12]).

5 Conclusion

We have presented a logic to reason about strategization in social network games. The logic is presented in two layers: local formulas talk about how a player strategizes based on threshold assumptions of strategies of neighbours. Global formulas assert strategy improvement by players and their reachability. We hope that the axiom system demonstrates that such local and global reasoning is sufficiently interesting.

Since we have treated outcomes as propositions, they come with a natural partial order: that of *implication*. However, $\neg p@i \land \langle i \rangle p@i$ implicitly asserts that p is a preferred outcome over $\neg p$ for player i. A natural question is to ask what utility functions and preference orderings can be expressed in such a logic (or enrichments thereof), and this leads in interesting directions.

The game equivalence presented here implicitly suggests an algebraic structure in the space of strategy profiles. But this is over strategic form games, which is quite different from that studied in [5], [14] over extensive form games. Moreover, though we have assumed a social relationship between players, the framework is non-cooperative, where players act individually. However social networks encompass both selfish and coalitional behaviour, and it would be interesting to study coalitional powers in social network games.

A very interesting question relates to mixed strategies. Note that the logic remains similar, and improvement graph dynamics can be studied. However, the strategy space is no longer discrete, and the transitive closure operator (based on finite paths) does not have much purchase. The basic modality is much more complicated and convergence issues are challenging.

While we have presented a Hilbert style axiom system, for reasoning about games it would be better to work with *sequents* of the form $\Gamma \vdash \phi$ where Γ is a theory, with formulas describing the game, thus constraining the space of profiles and ϕ is logically entailed. Developing such a proof theory seems to offer interesting challenges.

Acknowledgement: We thank Sunil Simon for discussions on social network games and Anantha Padmanabha for discussions on bisimulation classes of games. We thank the anonymous reviewers for helpful comments.

References

[1] Bonanno, G., *The logic of rational play in games of perfect information*, Economics and Philosophy **7** (1991), pp. 37–65.

[2] Chatterjee, K., T. A. Henzinger and N. Piterman, *Strategy logic*, Inf. Comput. **208** (2010), pp. 677–693.

[3] Clifford, P. and A. Sudbury, *A model for spatial conflict*, Biometrika **60** (1973), pp. 581–588.
    URL https://doi.org/10.1093/biomet/60.3.581

[4] Das, R., R. Ramanujam and S. Simon, *Reasoning about social choice and games in monadic fixed-point logic*, in: L. S. Moss, editor, *Proceedings Seventeenth Conference on Theoretical Aspects of Rationality and Knowledge, TARK 2019, Toulouse, France,*

17-19 July 2019, EPTCS **297**, 2019, pp. 106–120.
URL https://doi.org/10.4204/EPTCS.297.8

[5] Goranko, V., *The basic algebra of game equivalences*, Studia Logica **75** (2003), pp. 221–238.

[6] Harrenstein, P., W. van der Hoek, J.-J. Meyer and C. Witteven, *A modal characterization of nash equilibrium*, Fundamenta Informaticae **57:2**–4 (2003), pp. 281–321.

[7] Holley, R. A. and T. M. Liggett, *Ergodic Theorems for Weakly Interacting Infinite Systems and the Voter Model*, The Annals of Probability **3** (1975), pp. 643 – 663.
URL https://doi.org/10.1214/aop/1176996306

[8] Jackson, M. O. and Y. Zenou, *Games on networks*, in: *Handbook of Game Theory with Economic Applications*, 1 4, Elsevier, 2015 pp. 95–163.

[9] Mukund, M. and M. Nielsen, *Ccs, location and asynchronous transition systems*, in: R. K. Shyamasundar, editor, *Foundations of Software Technology and Theoretical Computer Science, 12th Conference, New Delhi, India, December 18-20, 1992, Proceedings*, Lecture Notes in Computer Science **652** (1992), pp. 328–341.
URL https://doi.org/10.1007/3-540-56287-7_116

[10] Ramanujam, R. and S. Simon, *A logical structure for strategies*, in: *Logic and the Foundations of Game and Decision Theory (LOFT 7)*, Texts in Logic and Games **3**, Amsterdam University Press, 2008 pp. 183–208.

[11] Seligman, J., F. Liu and P. Girard, *Facebook and the epistemic logic of friendship*, in: B. C. Schipper, editor, *Proceedings of the 14th TARK Conference, Chennai, India, January 7-9, 2013*.

[12] Simon, S. and K. R. Apt, *Social network games*, Journal of Logic and Computation **25** (2015), pp. 207–242.

[13] van Benthem, J., "Logic in Games," The MIT Press, 2014.

[14] van Benthem, J., N. Bezhanishvili and S. Enqvist, *A new game equivalence, its logic and algebra*, J. Philos. Log. **48** (2019), pp. 649–684.
URL https://doi.org/10.1007/s10992-018-9489-7

[15] van Benthem, J., S. Ghosh and R. Verbrugge, editors, "Models of Strategic Reasoning - Logics, Games, and Communities," Lecture Notes in Computer Science **8972**, Springer, 2015.
URL https://doi.org/10.1007/978-3-662-48540-8

[16] van Benthem, J., E. Pacuit and O. Roy, *Toward a theory of play: A logical perspective on games and interaction*, Games **2** (2011), pp. 52–86.
URL https://doi.org/10.3390/g2010052

New-Generation AIs
Reasoning about Norms and Values

Subproject Description
for the Logic for New-Generation AI project

Réka Markovich[a] Amro Najjar[a] Leendert van der Torre[a,b]

[a] *University of Luxembourg*

[b] *Zhejiang University*

Abstract

The recent rapid evolution of artificial intelligence and its widespread application in a multitude of domains led to the emergence of new heterogeneous systems where humans cohabit with software agents. The logics needed in these new-generation systems with intelligent behavior (AIs) have to accommodate reasoning about norms and values: these considerations drive humans' everyday life and decisions, we expect artificial intelligence tools to operate in our society taking these considerations very much into account. Some of these norms, like the legal ones, are mostly explicit, some, however, just like the values behind them, are not: the moral, social and cultural norms, while crucially affecting what people do and why, are usually not written, especially not precisely phrased, sometimes even not consciously reflected. In order to make the logics used for normative reasoning in artificial intelligence legitimate, next to the mathematical and technological requirements of such a formal system, we have to gather insights about these tacit or vague norms and values people reason with. In order to do so, next to the usual theoretical methodologies of developing new formal systems, data-driven and experimental methodologies with people, agents and texts are also needed revealing both the norms and values, and the ways people reason with them in different cultures. Engaging in this wide range of methodologies realizing a cross-disciplinary endeavor serves the purpose of gaining a comprehensive overview of what reasoning with norms and values is suitable for the new generation of AIs.

Keywords: normative reasoning, deontic logic, reasoning with values, data-driven approach, cross-cultural approach, cross-disciplinarity

1   Introduction

'Artificial intelligence' is used to describe the field studying and engineering intelligent agents [36], and nowadays it used as well as a synonym for intelligent agents themselves, either software agents or physical robots. In the new generation AIs, computer vision and machine learning play prominent roles, while at the same time being capable to represent knowledge and reason about it. Moreover, these new generation AIs are characterized by natural interaction with humans, and having significant social impact in the real world.

There are some key requirements that have been identified for these new generation AIs. One is that the natural interaction with humans implies that they can explain their behavior in an intelligent way to the humans they interact with, as well as to other AIs [4,22]. Another requirement, also implied by operating within our society interacting with humans, is that their actions be within the same normative framework as that of the humans'. Legal, moral and cultural norms and values provide the framework for our society defining the normative space of our actions, we expect the AIs' actions also to be within it, and their ability to reason with these norms and values to be a constituent of their intelligence [26]. In order to make sure that the decisions made by AIs are equipped with normative considerations, it is necessary to obtain and represent the existing legal and ethical norms of our societies. Achieving this goal requires facing and overcoming several challenges, though. While legal norms are mostly written, moral ones are almost never, the values and preferences over them are often not even consciously reflected. Thus, a methodology should exist for extracting norms, values and preferences from texts and from people's judgements and behavior.

Next to the norms themselves, preferences over these norms and values construct the basis for normative reasoning. The task of reasoning about values requires a parallel and highly connected endeavor when thinking about norms. Value and preference are everywhere in our daily life, and the notions of value and preference play key roles in research fields such as moral philosophy, and psychology.

A resilient reasoning system should be non-monotonic to handle conflicts between norms and values and should be able to reason on the meta-level too where more than one normative systems could be applied [25]. Also, such a reasoning system's accuracy should be verifiable. The technologies of cognition-driven and data-driven to establish a new model of value and normative reasoning research, and to combine value and ethical principle mining technology with the data-driven model in the field of legal reasoning. Moreover, we can model how an expected output, such as a set of ethical principles or a value preference, can be enforced given a multi-agent normative system and to investigate the computation complexity of the model.

The norms and values are different in the different societies [7], within those, they differ too in the different communities, and different stakeholders of a given AI tool might have different normative considerations even within the same community. Investigating and taking these differences into account while developing reasoning systems leads to a comparative study as an output of such a research project. This study is on similarities and differences of norms and values in the context of multi-culture and compare the ethical principles and normative systems between China and Europe, resulting from collecting data and cases of artificial intelligence related to norms and ethics in China and Europe, and evaluating and improving the algorithms of the models with instantiated data.

In the LNGAI project we adopt a non-monotonic logic to establish a value and normative reasoning model. We establish a data-driven ethical decision-making model. We also establish a multi-agent system based on norms and values and verify the accuracy of the system. Finally, in the LNGAI project we make a comparative study on similarities and differences of norms and values in artificial intelligence between China and European countries. This paper's layout is aligned to this research plan. During the discussion, we start from the following rough definitions:

Def 1. AI(s) = system(s) with intelligent behavior.

Def 2. New generation AI(s) = AI(s) (1) based on perception, representation and reasoning, and learning, (2) displaying natural interaction, and (3) having social impact.

Def 3. Explainable AI(s) = AI(s) that can explain behavior in an intelligent way to humans and other AIs.

Def 4. Legal, ethical, moral, social, cultural norms, normative system for AIs = individual and collective expressions of what is usual, typical or standard from the perspective of some discipline, institutions, organization, society or culture.

Def 5. Value, preference for AIs = individual and collective judgments of what is important in life.

Def 6. Deontic logic, preference logic for AIs = formal languages that can be used for the logical analysis of normative AIs, describing logical relations between the AIs and their rights and duties

2 Logic for Values and Norms

The prominent group of formalisms in normative reasoning is deontic logic. The foundations of deontic logic was created in the 1950's [43], embedded in the modal logic tradition, and its later variant is still referred to as standard deontic logic [15]. In this tradition, the main emphasis is on the obligations and permissions and the basis of the semantics are possible worlds. These systems, while being rather intuitive, suffer from paradoxes, and fundamentally monotonic. In the last decades, another tradition emerged in normative reasoning, closer to a rule-based approach, which explicitly refers to the norms themselves [3,29,33]. Agent-based modelling, for instance, BDI (belief-desire-intention) [34] can also be extended with the notion of obligation resulting in BOID models [12].

In a dynamic and open environment, the normative system and the agent's value system are unknown and changing. Also, normative systems often operate with exceptions and has to handle possible conflicts within the system. Not surprisingly, non-monotonic logics emerging in the 1980's in the filed of computer science [37] was subsequently applied to normative reasoning in the form of logic programming [5,39] and default logic [35,18]. But these formal languages based on computer programming are lacking in expressing ability and portraying diverse environments. In recent years, the formalism of argumentation theory and causal reasoning in norm and value reasoning has gradually attracted attention [9,24]. However, this work is still relatively preliminary. Thus, how to use the latest theories and methods of non-monotonic logic, formal argumentation systems and causal reasoning to carry out norms and values in an open and dynamic environment for reasoning with norms and values has not yet been systematically studied.

The so-called LogiKEy: Logic and Knowledge Engineering Framework and Methodology is also a recent development for designing and engineering ethical/legal reasoners, normative theories and deontic logics [10]. LogiKEy provides an integrative framework and methodology for using and developing new logics for reasoning with norms making it possible to continuously check the results against the ethical or legal theory we start from. Thus, LogiKEy is an essential asset when thinking about how the new generation AIs should reason about norms. It was created from an application-aiming approach, and it has the potential to revolutionize the area of deontic logic by addressing directly the decades-long challenge of how deontic logics and normative theories can be used in computer science applications. In the meantime, though, LogiKEy doesn't shift the focus from the theoretical basics: its pivotal property is the overarching nature. The unifying formal framework LogiKEy offers is based on semantical embeddings of deontic logics, logic combinations and ethical or legal domain theories in expressive classical higher-order logic (HOL). This meta-logical approach enables the provision of powerful tool support in LogiKEy: off-the-shelf theorem provers and model finders for HOL are assisting the LogiKEy designer of ethical intelligent agents to flexibly experiment with underlying logics and their combinations, with ethical or legal domain theories, and with concrete examples at the same time.

The task of reasoning about values requires a paralel and highly connected endeavor when thinking about norms. Value and preference are everywhere in our daily life, and the notions of value and preference play key roles in research fields such as moral philosophy, psychology, decision theory, game theory, and social choice. Value is often expressed as monadic preference, such as "I prefer to go to the West Lake", "the committee prefers to make its decisions available on the website", "it is preferred to be honest", or "people prefer symmetric faces". Preference is usually expressed in terms of a comparison between two objects or situations, using comparative statements such as "If we are served duck, then I prefer rice over noodles", or "peace is preferred over war".

Preference logic formalises reasoning about value and preference statements. Reasoning about preference is challenging, both conceptually and computationally. Some of the conceptual challenges are the aggregation of preferences, the change of preference, or the definition of the ceteris paribus proviso. Examples of computational challenges are efficient querying of preference, preference elicitation, communication of preference, and non-monotonic reasoning about preference.

Traditionally the emphasis in reasoning about value and preference was on intrinsic preference, but more recently the emphasis has shifted to extrinsic preference. Roughly, extrinsic preferences are based on reasons, while intrinsic preferences are not. For example, Von Wright's logic of preference [46] is explicitly restricted to intrinsic preferences in the sense that he considers preferences not having an extrinsic reason or motivation, whereas in deontic logic the preference ordering is derived from a set of norms in norm-based semantics [28]. In general, the logic of extrinsic preference makes the reasons explicit, and thus reasons about both preference and the reasons for preference. In moral reasoning, value may be one of the reasons behind norms.

As a consequence of making the reasons for extrinsic preference explicit, the formalisation of value and preference change has become a central concern for preference logic as well. For example, extrinsic preferences may be changed by commanding or promulgating norms. In general, extrinsic preferences change when the reasons change, or when the priorities amongst these reasons change, whereas intrinsic preference cannot change in the same way. Contextual or intrinsic preference may change due to changing beliefs. A person may prefer to pursue an academic career, but adjust his preference if he learns about the consequences of being a professor.

Finally, preference often comes with a ceteris paribus proviso, which refers to the condition of "other things being equal". Therefore, in preference logic special attention is given to ceteris paribus preference. Moreover, to formalize the use of value and preference in practical reasoning, the logic of preference needs to be developed further, for instance, combined with the logic of belief.

3   Data-driven Ethical Decision-making

With the rise of complex AI systems and the advancements of their autonomy, the issues of ethical decision-making is receiving growing attention since, as it the case with human decision makers, these systems can be confronted with critical situations in which the decisions to be made can have heavy ethical consequences. These type of situations are also knows as moral dilemmas in which, for example, a self-driving car needs to take a decision in a critical situation where human life is at stake. Traditionally, the normative systems and deontological logics governing the behavior of these proposed systems relied on complex ethical principles and policies that should be formally specified [16,8]. Nevertheless, the problem of endowing AI system with the capacity to make ethical decisions has remain a challenging tasks for the past years [44]. This issue is further complicated by the fact that ethical principles are often dynamic, across cultures, geographies, as well as other human related factors (e.g., gender). These differences were elegantly highlighted by the Moral Machine experiment [7].

The recent breakthroughs in big-data and machine learning made their way into AI ethics and novel data-driven approaches have shown remarkable preliminary results promising to tackle practical solutions. In this context, data are not only used evaluate the outcomes, but are also used to produce new inference models. More specifically, in many cases, it is difficult to rely on formal specifications to reflect the actual decision making situation as implied by the real-data sensed by the agent. Instead, machine learning mechanisms are used to mine implicit reasoning patterns in the data and use them to predict new cases. The quantification and modeling of such norms and values can be performed via machine learning and parametric techniques using Inverse Reinforcement Learning (IRL) [1], imitation learning [42], inverse game theory [45], and norm inference [20].

Despite these recent advances, many challenges persist:

C1 **Data-hungry algorithms**: Most of off-the-shelf machine learning mechanisms require big training datasets. Thus, when applied for Data-driven ethical decision-making, these mechanism would require training datasets to be representative of societal choices and ethical values [32].

C2 **Human-Trained Machine Learning**: Human behavior and performance should provide the baseline to teach the AI and benchmark its behavior against humans. However, most of existing works fail to define and measure human operator performance in real-time context [8]. Moreover, in some applications, value and its ranking are the tacit knowledge possessed by human participants. Contained in the data related to the subject, it is difficult to express in a formal way, and it is also difficult to obtain and process.

C3 **Context-dependence**: Ethical decision-making is mostly context-dependent. Thus, compared to the often brittle traditional approaches, data-driven mechanisms are easier to adapt to dynamically changing scenarios. However, this necessitates that training data are representative of the different contexts.

C4 **Black-box Machine Learning Mechanisms**: most of machine learning mechanisms used for data-driven ethical decision-making are black-box mechanisms whose inner-workings are subsymbolic and, therefore, non-understandable by humans.

4   Norm- and Value-Based Multiagent Systems

In dynamic, open and heterogeneous societies, agents are expected to be collaborating with other agents (human and artificial agents). This is the case, for instance, of self-driving cars, medical robots, and home service robots who all demonstrate autonomous capabilities. In these application scenarios, there are often interactions and collaborations between multiple intelligent agents, in the emerging, dynamic, open and multi-agent system, the autonomous behavior and decision-making of each intelligent agent may have a huge impact on society. If left unchecked, self-interested agents my cause harms to others while trying to seek their goals. Similar to their roles in human societies, norms provide a means to regulate agent behavior [27] and make them conform to certain social expectations at the ethical and legal levels. Thus, the results of the face-to-face output conform to certain social expectations at the ethical level, such as: the need to add norms to the system to guide the overall behavior of the system, and to monitor and control the behavior of intelligent agents.

Nevertheless, developing normative multi-agent system raises a specific set of challenges:

C1 **Norms in Open vs Closed Agent Societies**: Often, traditional normative systems are designed for closed societies whose value system becomes the basis for the normative system. The rules stipulated reflect the general importance of society. But when the environment in the society is open, which is the case of most of multi-agent systems, we cannot guarantee that the individual agents entering the society have the same value system as the society. (e.g., when a self-driving car enters a new traffic environment). Detecting and resolving this conflict arising between the two normative systems and ensuring that individual agents have compatible value systems in order to ensure that the overall behavior of the society meets our expectations are open challenges.

C2 **Legalistic vs interactionist view of norms**: The former considers that the normative agent system as a regulatory instrument regulating the emerging behavior of open systems without enforcing the desired behavior. In such a case, agents are often motivated by sanctions to stick to norms, rather than by their sharing of the norms, whereas the later (the interactionist view) is a bottom-up approach in which norms can be seen, as regularities of behavior which emerge without any enforcement system because agents conform to them either because (i) their goals happen to coincide (ii) because they feel themselves as part of the group (iii) or because they share the same values of other agent. In this case, sanctions become not always necessary even for norm violation [11,13].

C3 **Subjective and Cross-cultural differences**: as has been shown by recent user studies [7,21], norms tend to be dependent on factors such as culture (i.e., European vs east Asian), to be context-dependent and application dependent, and to be different from one user to another. In order to cope with this challenge, it is necessary to study the relationship between the subject's decision-making based on the value system and norms and the overall behavior of the system.

5   Evaluation in Multi-culture

The existing research on norms and value reasoning is generally based on a specific geographical or cultural background [23,6]. But norms and values greatly differ in and of the different countries, societies, communities, and individuals. This aspect has to be taken account when developing formal systems of reasoning with norms and values.

Law is a geographically-socially determined system of norms: the legal systems—both on the level of (constitutional) values and the very norms based on the former—are different in the different countries. These dissimilarities are especially strong between countries with different histories, cultural and political backgrounds. And while it would be rather difficult to take all countries' all norms and values into account, aiming for a cross-cultural reference and considering some relevant differences between China and the European countries is definitely feasible and needed for a comprehensive approach of an adequate normative reasoning within AI.

One might expect that ethics is less divisive than law, but apparently this is far from sure. The well-known and often discussed Moral Machine[7] is a cross-cultural study of ethical norms and values based on self-driving car scenarios. While there are many critical opinions regarding its methodology, the simulation[8], and the questions used in the experiment, what the study apparently clearly shows is that there are no globally valid ethical considerations, moral norms. The Moral Machine experiment and study concerned a specific environment of self-driving cars with different scenarios realizing variants of the so-called Trolley Problem[14], so one might think that one a higher, or more abstract level, the values and norms might converge. But an investigation of 84 newly written ethical guidelines it was found that while there are some emerging values, "no single ethical principle appeared to be common to the entire corpus of documents".[19] Next to the object level of—specific or abstract—norms, cultural and historical differences might affect the meta-level considerations too: what source should be accepted as a source of valid norms, who can and should decide about what norms should drive the mechanisms of, for example, autonomus systems.

This takes us to different tasks when developing formal models of reasoning with norms and values for new generation AI. One one hand, we need to take these differences into account when considering and collecting the norms and values themselves. There might be differences also in the way people reason with the norms and values in the different cultures and these possible differences have to accommodated too, but first data need to be gained on the existing dissimilarities. On the top of it, we need a resilient reasoning system which, optimally, is not only applicable for different ways of reasoning with norms and values, but also applicable in situations where meta-reasoning is needed about what norms and values should be applied in the reasoning. On the other hand, these differences have to be taken into consideration when verifying the reasoning system: it has to be checked against different benchmark examples coming from different cultures and evaluated in different environments. Collecting data as input for developing an adequate reasoning system and the results of its evaluation against the different cultures and their norms and values will provide an extensive comparative study between China and the European countries.

For taking this aspect seriously, one needs to establish methods for collecting data from different cultures and for evaluating the adaptability of the formal model in a cross-cultural context to answer the question: how does the model perform in different geographical and cultural contexts?

To do so, the we identify the following steps:

(i) Devise new metrics allowing to measure how adaptable a formal model to different cultures, and how data can be collected in an invasive-less manner. A similar endeavor is being conducted in the domain of explainable AI where new metrics are being defined [17] to measure the satisfaction and trustworthiness inspired by an explanation. In addition, a distinction is made between objective and subjective understandability, thereby giving room for personal and perhaps inter-cultural differences with regards to explaination reception and understandability.

(ii) Psychometric scales and statistical significance: when conducting user studies, adequate answer scales should be selected. Moreover, the statistical significance of the outcomes should be tested. One good option is the Likert scale [2]. The latter is commonly used in research and surveys to measure attitude, providing a range of responses to a given question or statement. The typical Likert scale is a 5- or a 7-point ordinal scale used by participants to rate the degree to which they agree or disagree with a question or a statement. While the Likert scale is widely used in scientific research, there has been a long-standing controversy regarding the analysis of ordinal data [41]. In fact, analyzing the outcomes of the Likert scale, and the use of parametric tests to analyze ordinal data in general, has been subject to an active and ongoing debate. In order to adequately obtain the user answers, the psychometric scales to be used should be selected, and the methods used to establish the statistical significance should be updated accordingly.

6   Towards Reasoning with Norms and Values in New Generation AIs All Around the Globe

Norms and values drive humans' everyday life and decisions. Since new-generation AIs operate in our society and cohabit with humans, we expect these new generation AIs to take norms and values into account, and their logics have to accommodate reasoning about these norms and values. Despite the recent move to discuss AI from interdisciplinary standpoint covering ethical, legal and societal aspects, most of this engagement originates from Euro-American scholars with obvious influences from the Western epistemic tradition. This results in the marginalization of non-western knowledge systems in the study of AI ethics.

In order to make AI acceptable for global audience, several barriers and centrisms needs to be overcame and a more inclusive approach involving east Asia, but also Africa and the middle-east should be devised. The resulting intercultural approach to the ethics of AI should inform the formation of policies and guidelines to regulate the design and use of AI. This research direction has been recommended by several recent initiatives from IEEE[40] and UNESCO advocating a global instrument on the ethics of AI, which would also serve as guidelines for practitioners, governments and policy-makers.

These differences, profoundly shape the nature of contributions to the field on issues of data privacy, social robotics, conceptions of artificial moral agency, moral status and patiency, autonomous weapons systems, big data and the likes. For instance, as noted by Metz [30,31], there are recurrent salient features that can be found in many sub-Saharan cultures that are not found (in the same way) elsewhere in the world. This does not mean these features cannot be found in other cultures, it just means that they are more recurrent in Africa. The same can be said of Western, Middle Eastern and South-East Asian cultures[38]. Hence, both Afro-ethical (e.g. Ubuntu traditions) and Confucian ethical systems share similarities since both are collectivists systems whose normative principles rest heavily on a collectivist disposition notably when it comes to determine right or wrong action. In contrast to Western ideals, built on advancing individualism following the age of enlightenment and the values spread by the industrial revolution, Afro-ethical and Confucian moral values share principles that advance collective progress, harmony and group cohesion.

Also, to accommodate the clearly existing differences between the different cultures, we have to gather insights also about the tacit or vague norms and values people reason with in order to make the logics used for normative reasoning in artificial intelligence legitimate. In order to do so, next to the usual theoretical methodologies of developing new formal systems, we need to engage with data-driven and experimental methodologies with people, agents and texts, in order to reveal both the norms and values, and the ways people reason with them in different cultures. Employing in this wide range of methodologies realizing a cross-disciplinary endeavor serves the purpose of gaining a comprehensive overview of what reasoning with norms and values is suitable for the new generation of AIs.

References

[1] Abel, D., J. MacGlashan and M. L. Littman, *Reinforcement learning as a framework for ethical decision making.*, , **16**, Phoenix, AZ, 2016, p. 02.

[2] Albaum, G., *The likert scale revisited*, Market Research Society. Journal. **39** (1997), pp. 1–21.

[3] Alchourrón, C. E. and E. Bulygin, "Normative Systems," Springer-Verlag New York – Wien, 1971.

[4] Anjomshoae, S., A. Najjar, D. Calvaresi and K. Främling, *Explainable agents and robots: Results from a systematic literature review*, in: *18th International Conference on Autonomous Agents and Multiagent Systems (AAMAS 2019), Montreal, Canada, May 13–17, 2019*, International Foundation for Autonomous Agents and Multiagent Systems, 2019, pp. 1078–1088.

[5] APT, K. R., *Chapter 10 - logic programming*, in: J. VAN LEEUWEN, editor, *Formal Models and Semantics*, Handbook of Theoretical Computer Science, Elsevier, Amsterdam, 1990 pp. 493–574.
URL https://www.sciencedirect.com/science/article/pii/B9780444880741500159

[6] Awad, E., M. Anderson, S. L. Anderson and B. Liao, *An approach for combining ethical principles with public opinion to guide public policy*, Artif. Intell. **287** (2020), p. 103349.
URL https://doi.org/10.1016/j.artint.2020.103349

[7] Awad, E., S. Dsouza, R. Kim, J. Schulz, J. Henrich, A. Shariff, J.-F. Bonnefon and I. Rahwan, *The moral machine experiment*, Nature **563** (2018), pp. 59–64.

[8] Behzadan, V., J. Minton and A. Munir, *Trolleymod v1. 0: An open-source simulation and data-collection platform for ethical decision making in autonomous vehicles*, in: *Proceedings of the 2019 AAAI/ACM Conference on AI, Ethics, and Society*, 2019, pp. 391–395.

[9] Bench-Capon, T. and S. Modgil, *Norms and value based reasoning: Justifying compliance and violation*, Artificial Intelligence and Law **25** (2017), pp. 29–64.

[10] Benzmüller, C., X. Parent and L. van der Torre, *Designing normative theories for ethical and legal reasoning: Logikey framework, methodology, and tool support*, Artificial Intelligence **287** (2020), p. 103348.
URL https://www.sciencedirect.com/science/article/pii/S0004370219301110

[11] Boella, G., L. Van Der Torre and H. Verhagen, *Ten challenges for normative multiagent systems*, in: *Dagstuhl Seminar Proceedings*, Schloss Dagstuhl-Leibniz-Zentrum für Informatik, 2008.

[12] Broersen, J., M. Dastani, J. Hulstijn, Z. Huang and L. van der Torre, *The boid architecture - conflicts between beliefs, obligations, intentions and desires*, in: *In Proceedings of the Fifth International Conference on Autonomous Agents* (2001), pp. 9–16.

[13] Castelfranchi, C., *Modelling social action for ai agents*, Artificial intelligence **103** (1998), pp. 157–182.

[14] Foot, P., *The problem of abortion and the doctrine of the double effect*, Oxford Review **5** (1967), pp. 5–15.

[15] Gabbay, D., J. Horty, X. Parent, R. van der Meyden and L. van der Torre, editors, "Handbook of Deontic Logic and Normative Systems," College Publications, 2013.

[16] Greene, J., F. Rossi, J. Tasioulas, K. Venable and B. Williams, *Embedding ethical principles in collective decision support systems*, , **30**, 2016.

[17] Hoffman, R. R., S. T. Mueller, G. Klein and J. Litman, *Metrics for explainable ai: Challenges and prospects*, arXiv preprint arXiv:1812.04608 (2018).

[18] Horty, J. F., "Reasons as Defaults," Oxford University Press, 2012.

[19] Jobin, A., M. Ienca and E. Vayena, *The Global Landscape of AI Ethics Guidelines.*, Nature Machine Intelligence **1** (2019), pp. 389–399.

[20] Kasenberg, D., T. Arnold and M. Scheutz, *Norms, rewards, and the intentional stance: Comparing machine learning approaches to ethical training*, in: *Proceedings of the 2018 AAAI/ACM Conference on AI, Ethics, and Society*, 2018, pp. 184–190.

[21] Komatsu, T., B. F. Malle and M. Scheutz, *Blaming the reluctant robot: parallel blame judgments for robots in moral dilemmas across us and japan*, in: *Proceedings of the 2021 ACM/IEEE International Conference on Human-Robot Interaction*, 2021, pp. 63–72.

[22] Langley, P., B. Meadows, M. Sridharan and D. Choi, *Explainable agency for intelligent autonomous systems.*, , **17**, 2017, pp. 4762–4763.

[23] Liao, B., M. Anderson and S. L. Anderson, *Representation, justification, and explanation in a value-driven agent: an argumentation-based approach*, AI Ethics **1** (2021), pp. 5–19.
URL https://doi.org/10.1007/s43681-020-00001-8

[24] Liao, B., N. Oren, L. Van Der Torre and S. Villata, *Prioritized norms in formal argumentation*, Journal of Logic and Computation **29** (2019), pp. 215–240.
URL https://hal.archives-ouvertes.fr/hal-02381116

[25] Liao, B., M. Slavkovik and L. van der Torre, *Building Jiminy Cricket: An Architecture for Moral Agreements Among Stakeholders*, in: *Proceedings of the 2019 AAAI/ACM Artificial Intelligence, Ethics and Society* (2019), pp. 147–153.

[26] Luck, M., S. Mahmoud, F. Meneguzzi, M. Kollingbaum, T. J. Norman, N. Criado and M. S. Fagundes, *Normative agents*, in: *Agreement technologies*, Springer, 2013 pp. 209–220.

[27] Luck, M., S. Mahmoud, F. Meneguzzi, M. Kollingbaum, T. J. Norman, N. Criado and M. S. Fagundes, "Normative Agents," Springer Netherlands, Dordrecht, 2013 pp. 209–220.
URL https://doi.org/10.1007/978-94-007-5583-3_14

[28] Makinson, D., *On a fundamental problem of deontic logic*, in: *Norms, Logics and Information Systems. New Studies in Deontic Logic and Computer Science*, IOS press, 1999 pp. 29–53.

[29] Makinson, D. and L. van der Torre, *Input/output logics*, Journal of Philosophical Logic **29** (2000), pp. 383–408.

[30] Metz, T., *Toward an african moral theory*, Journal of Political Philosophy **15** (2007), pp. 321–341.

[31] Metz, T., *Toward an african moral theory (revised edition)*, in: *Themes, issues and problems in African philosophy*, Springer, 2017 pp. 97–119.

[32] Noothigattu, R., S. Gaikwad, E. Awad, S. Dsouza, I. Rahwan, P. Ravikumar and A. Procaccia, *A voting-based system for ethical decision making*, , **32**, 2018.

[33] Parent, X. and L. van der Torre, *Input/output logic*, in: D. Gabbay, J. Horty, X. Parent, R. van der Meyden and L. van der Torre, editors, *Handbook of Deontic Logic and Normative Systems*, College Publications, 2013 pp. 353–406.

[34] Rao, A. S. and M. P. Georgeff, *Bdi agents: From theory to practice*, in: *IN PROCEEDINGS OF THE FIRST INTERNATIONAL CONFERENCE ON MULTI-AGENT SYSTEMS (ICMAS-95*, 1995, pp. 312–319.

[35] Reiter, R., *A logic for default reasoning*, Artificial Intelligence **13** (1980), p. 81–132.

[36] Russell, S. J. and P. Norvig, *Artificial intelligence: a modern approach* (2010).

[37] Schlechta, K., "Nonmonotonic Logics," Springer, 1997.

[38] Segun, S. T., *Critically engaging the ethics of ai for a global audience*, Ethics and Information Technology (2020), pp. 1–7.

[39] Sergot, M. J., F. Sadri, R. A. Kowalski, F. Kriwaczek, P. Hammond and H. T. Cory, *The british nationality act as a logic program*, Commun. ACM **29** (1986), p. 370–386. URL https://doi.org/10.1145/5689.5920

[40] Shahriari, K. and M. Shahriari, *Ieee standard review—ethically aligned design: A vision for prioritizing human wellbeing with artificial intelligence and autonomous systems*, in: *2017 IEEE Canada International Humanitarian Technology Conference (IHTC)*, IEEE, 2017, pp. 197–201.

[41] Sullivan, G. M. and A. R. Artino Jr, *Analyzing and interpreting data from likert-type scales*, Journal of graduate medical education **5** (2013), p. 541.

[42] Taylor, J., E. Yudkowsky, P. LaVictoire and A. Critch, *Alignment for advanced machine learning systems*, Ethics of Artificial Intelligence (2016), pp. 342–382.

[43] Von Wright, G., *Deontic logic*, Mind : a Quarterly Review of Psychology and Philosophy **60** (1951). URL http://search.proquest.com/docview/1293708393/

[44] Wallach, W. and C. Allen, "Moral machines: Teaching robots right from wrong," Oxford University Press, 2008.

[45] Wang, Y., Y. Wan and Z. Wang, *Using experimental game theory to transit human values to ethical ai*, arXiv preprint arXiv:1711.05905 (2017).

[46] Wright, G. H. v. G. H., "Norm and Action : a Logical Enquiry," International library of philosophy and scientific method, Routledge and Kegan Paul, London, 1963.

Getting Consensus Through a Context-Based Argumentation Framework

Zhe Yu, Shier Ju [1]

Institute of Logic and Cognition, Department of Philosophy, Sun Yat-sen University
Guangzhou 510275, China

Abstract

Argumentation must take place within some contexts, which contains particular norms and values. This paper introduces a revised version of the Context-based Argumentation Framework proposed in a previous paper. The argumentation theory is mainly built upon $ASPIC^+$. Meanwhile, ideas behind the definition of contexts and consensus are inspired by the Generalized Argumentation Theory. Compared with other work dealing with multi-agent reasoning based on formal argumentation, this paper emphasizes the dynamic feature of contexts and proposes an approach to obtain consensus that combines a pragmatic perspective.

Keywords: structured argumentation, contexts, values, consensus.

1   Introduction

People argue to resolve disputes, and one of the main purpose is to reach consensus on some key issues. These debates must take place in some certain contexts and involve particular social norms and values. Moreover, in many cases, for the purpose of persuasion, some participants may change the context as a strategy. For example, consider the following conversation between a mother and her child.

Example 1.1
Child:"Mom, I don't want to do my homework tonight, because today is Friday!"
Mother:"Just remember that you will not be allowed to play computer games or go out on the weekend until your homework is done."

In the above conversation, the mother changed the context by raising a new issue of "whether to have more fun on the weekend", in an attempt to make the child reconsider the priority between "having a rest" and "doing homework".

To model contextual information by formal argumentation, the initial idea of a Context-based Argumentation System (CAS) is introduced in a previous paper [29].

---

[1] Corresponding author.
Email addresses: yuzh28@mail.sysu.edu.cn (Z. Yu), hssjse@mail.sysu.edu.cn (S. Ju)
This research is supported by the National Social Science Foundation of China (No.20&ZD047), the Philosophy and Social Science Youth Projects of Guangdong Province (No.GD19CZX03) and the China Postdoctoral Science Foundation (No.2019M663353). The authors would like to thank the anonymous reviewers for their comments and suggestions.

132

Basically, we give a formal definition of (a sequence of) contexts containing norms, values and the priority orderings given by different participants, then merge these elements with the standard *ASPIC*$^+$ framework proposed by Modgil and Prakken [22].

The intuition behind the Context-based Argumentation System is in line with Perelman's idea about *agreement* and *values* [24]. In addition, it has been inspired by the *Generalized Argumentation Theory* [16,17,18], which clarifies that argumentation is "a series of discourse action with argumentative function that are produced in a dynamical context by participants belonging to one or more socio-cultural communities with norms of their own" [18].

In this paper, we present a revised version of CAS. In particular, in order to better depict a context, we re-examine and modify its definition; we specify the way to obtain preferences on the basis of priority orderings over values, as well as the requirements of a 'well-defined' CAS according to [22]; in addition, we will discuss what are 'proper' consensus and 'desired' consensus, then compare our approach with some related work.

The structure of this paper is as follows: Section 2 introduces the basic settings of CAS, as well as the definition of *consensus*, then illustrates our ideas through a case modelling. Section 3 compares CAS with some close methods and introduces other related work. At last, section 4 concludes this paper.

2   The Context-based Argumentation Framework and Properties

2.1   Definitions of the Revised CAS

According to the Generalized Argumentation Theory [18], contexts are changeable in the process of a dispute and can appear as a sequence. Different parties involved in the dispute exchange arguments based on social norms under certain cultural background. When the context is shifted, participants may change their opinions, then consensus on the controversial issues may be achieved.

Since contexts can constitute a sequence, we use a subscript i to denote the order of the context in a dialogue. According to the above basic idea, we present the following definition of contexts.

Definition 2.1 [Contexts] A context is a tuple $C_i = \langle I_i, N_i, V_i, val, P_i \rangle (i = 1, 2, \ldots, n)$, where

- I_i is a set of issues closed under negation ($\neg$); $\bigcup_{i=1}^{n} I_i$ is denoted by $\mathscr{I}_i$.

- N_i is a set of norms of the form $\varphi_1, \ldots, \varphi_n \xrightarrow{v_u} \varphi$ ('$\Longrightarrow$' denotes an uncertain inference and v_u denotes a basic value 'u' it associated with, while φ_i and φ are elements in the logical language of an argumentation theory); $\bigcup_{i=1}^{n} N_i$ is denoted by $\mathscr{N}_i$.

- $V_i = \{v_p, v_q, \ldots, v_z\}$ is a set of values ('$p, q, \ldots, z$' represent names of the values); $\bigcup_{i=1}^{n} V_i$ is denoted by $\mathscr{V}_i$[2].

- *val* is a function from elements of $\mathscr{N}_i$ to elements of $\mathscr{V}_i$.

---

[2]   In order to take into account all the norms and values that have appeared in the current context and in the previous contexts of the dialogue, we often use $\mathscr{N}_i$ or $\mathscr{V}_i$ instead of N_i or V_i in the following definitions of CAS.

- $P_i = \{\lesssim_1, \lesssim_2, \ldots, \lesssim_m\}$ is a set of preorderings over $\mathcal{V}_i$, where '$1, 2, \ldots, m$' denote various the parties that involved the dispute, we write: 1. $v_n \lesssim v_m$, iff v_m is at least as preferred as v_n; 2. $v_n < v_m$, iff $v_n \lesssim v_m$ and not $v_m \lesssim v_n$; 3. $v_n \approx v_m$, iff $v_n \gtrsim v_m$ and $v_m \gtrsim v_n$.

Intuitively, to change a context means to bring up new issues, norms and values, which may make the other participants reconsider their priority orderings over values. We focus on resolving disputes among the proponent and the opponent of some key issues. Note that, this does not mean there are only two parties, since various priority orderings over values can lead to the same conclusion. Accordingly, compared with the initial definition of contexts in [29], a set of *issues* (I) is added into every context, which contains controversial topics or important decisions that are under consideration within the dispute. Further, we modify the setting of norms by clarifying that the inferences are defeasible. The reason is that norms are closely relative to socio-cultural environment, therefore are comparable and defeasible in nature, and if there are any strict norms, they can be modelled by strict rules introduced in the following definition of CAS. Moreover, in the current paper we assume that each norm is associated with only one basic value it is based on or related to. Different norms may be associated with the same value.

A context-based argumentation system is built upon $ASPIC^+$ framework [22], while contextual information is included.

Definition 2.2 [Context-Based Argumentation System] An argumentation system based on contexts is a tuple $CAS = (\mathcal{L}, ^-, \mathcal{R}, \mathcal{C}, n)$, where:

- $\mathcal{L}$ is a logical language.
- $^-$ is a function from $\mathcal{L}$ to $2^{\mathcal{L}}$, s.t.: 1. φ is a contrary of ψ if $\varphi \in \overline{\psi}$ and $\psi \notin \overline{\varphi}$; 2. φ is a contradictory of ψ (denoted by '$\varphi = -\psi$'), if $\varphi \in \overline{\psi}$ and $\psi \in \overline{\varphi}$ [3]; 3. each $\varphi \in \mathcal{L}$ has at least one contradictory.
- $\mathcal{R} = \mathcal{R}_s \cup \mathcal{R}_d \cup \mathcal{N}_i$ is a set of strict ($\mathcal{R}_s$) and defeasible ($\mathcal{R}_d \cup \mathcal{N}_i$) inference rules of the form $\varphi_1, \ldots, \varphi_n \longrightarrow \varphi$ and $\varphi_1, \ldots, \varphi_n \overset{(v_u)}{\Longrightarrow} \varphi$ respectively (φ_i, φ are meta-variables ranging over $\mathcal{L}$); $\mathcal{R}_s \cap (\mathcal{R}_d \cup \mathcal{N}_i) = \emptyset$.
- $\mathcal{C} = \{C_1, C_2, \ldots, C_n\}$ is a set of contexts, which contains the sequence of contexts in the dialogue, where $C_i = \langle I_i, N_i, V_i, val, P_i \rangle$ ($i = 1, 2, \ldots, n$), s.t. $\mathcal{I}_i \cup \mathcal{V}_i \subseteq \mathcal{L}$.
- n is a naming function s.t. $n : \mathcal{R}_d \to \mathcal{L}$.

In order to include the previously existing rules, $\mathcal{R}$ contains all the norms that appear in the context C_i as well as the contexts before it (i.e., $\mathcal{N}_i$). A defeasible rule belongs to the set $\mathcal{N}_i$ if and only if it is associated with a value.

An argumentation system together with a knowledge base $\mathcal{K}$ compose an *argumentation theory* [22]. Elements in $\mathcal{K}$ serve as the premises of arguments, which are divided into *axioms* and *ordinary premises*, denoted by two disjoint sets $\mathcal{K}_n$ and $\mathcal{K}_p$ respectively, of which only the ordinary premises are attackable.

---

[3] For all $\varphi \in \mathcal{L}$, we have $\neg - \varphi \in \overline{\varphi}$ and for all $\neg \varphi \in \mathcal{L}$, we have $\varphi \in \overline{\neg \varphi}$.

Definition 2.3 [Context-based Argumentation Theory] Let $CAT = (CAS, \mathcal{K})$ be a context-based argumentation theory, where $CAS = (\mathcal{L}, ^-, \mathcal{R}, \mathcal{C}, n)$ is a context-based argumentation system and $\mathcal{K} \subseteq \mathcal{L}$ is a knowledge base s.t. $\mathcal{K} = \mathcal{K}_n \cup \mathcal{K}_p$ and $\mathcal{K}_n \cap \mathcal{K}_p = \emptyset$, $\mathcal{K}_n$ is the set of axioms and $\mathcal{K}_p$ is the set of ordinary premises.

Based on an argumentation theory, arguments can be constructed. Adapted from [22], we use $Prem(A)$ to denote the set of all the formulas of $\mathcal{K}$ used to build an argument A, $Conc(A)$ to denote the conclusion of A, $Sub(A)$ to denote the set of all the sub-arguments of A, and $DefRules(A)$ to denote the set of all the regular defeasible rules (rules in $\mathcal{R}_d$) applied in A. What is more, we use $Norms(A)$ to denote the set of all the norms in $\mathcal{N}_i$ applied in A, and $Values(A)$ to denote the set of all the values associated to norms in A. Arguments in a CAT is defined as follows.

Definition 2.4 [Arguments] An argument A on the basis of a $CAS = (\mathcal{L}, ^-, \mathcal{R}, \mathcal{C}, n)$ and $\mathcal{K}$ has one of the following forms:

(i) φ, if $\varphi \in \mathcal{K}$ with: $Prem(A) = \{\varphi\}$, $Conc(A) = \varphi$, $Sub(A) = \{\varphi\}$, $DefRules(A) = \emptyset$, $Norms(A) = \emptyset$, $Values(A) = \emptyset$;

(ii) $A_1, \ldots, A_n \longrightarrow / \overset{(v_u)}{\Longrightarrow} \psi$, if $A_1, \ldots, A_n$ $(n \geq 1)$ are arguments, s.t. there exists a strict rule/defeasible rule/norm $Conc(A_1), \ldots, Conc(A_n) \longrightarrow / \overset{(v_u)}{\Longrightarrow} \psi$ in $\mathcal{R}$ with: $Prem(A) = Prem(A_1) \cup \ldots \cup Prem(A_n)$; $Conc(A) = \psi$; $Sub(A) = Sub(A_1) \cup \ldots \cup Sub(A_n) \cup \{A\}$; $DefRules(A) = DefRules(A_1) \cup \ldots \cup DefRules(A_n)$ $(\cup\{Conc(A_1), \ldots, Conc(A_n) \Longrightarrow \psi\})$; $Norms(A) = Norms(A_1) \cup \ldots \cup Norms(A_n)$ $(\cup\{Conc(A_1), \ldots, Conc(A_n) \overset{v_u}{\Longrightarrow} \psi\})$; $Values(A) = Values(A_1) \cup \ldots \cup Values(A_n)$ $(\cup\{v_u\})$.

$ASPIC^+$ can model three kinds of conflicts between arguments, adapted from [22], the attack relation is defined as follows.

Definition 2.5 [Attack] An argument A attacks argument B, iff A undercuts, rebuts or undermines B, where: 1. A **undercuts** B on B', iff $B' \in Sub(B)$ s.t. $TopRule(B') = r \in \mathcal{R}_d$ and $Conc(A) \in \overline{n(r)}$ ('$n(r)$' means that rule r is applicable); 2. A **rebuts** B on B', iff $Conc(A) \in \overline{\varphi}$ for some $B' \in Sub(B)$ of the form $B''_1, \ldots, B''_n \overset{(v_u)}{\Longrightarrow} \varphi$; A **contrary-rebuts** B iff $Conc(A)$ is a contrary of φ. 3. A **undermines** B on B', iff $B' = \varphi$ and $\varphi \in Prem(B) \cap \mathcal{K}_p$, s.t. $Conc(A) \in \overline{\varphi}$; A **contrary-undermines** B iff $Conc(A)$ is a contrary of φ.

Except for the undercutting, contrary-rebutting and contrary-undermining [4], whether an attack can succeed as a *defeat* relation depends on a preferences ordering $\preceq$ on the set of arguments, which can be 'lifted' based on two preorderings on the set of ordinary premises $\mathcal{K}_p$ and defeasible rules $\mathcal{R}_d$ according to the *last-link* or *weakest-link* principles [22].

To apply the two principles, two set comparison approach can be adopted, i.e. the *Elitist* and *Democratic* approaches [11]. Let $s \in \{Eli, Dem\}$, the two approaches can be defined as follows.

---

[4] These kinds of attacks are 'preference-independent' according to [22].

Definition 2.6 [Set Comparison] Let Γ and Γ' be two finite sets, $\unlhd_s$ denotes a set comparison: 1. if $\Gamma = \emptyset$ then $\Gamma \ntrianglelefteq_s \Gamma'$; 2. if $\Gamma' = \emptyset$ and $\Gamma \neq \emptyset$, then $\Gamma \unlhd_s \Gamma'$; 3. for a preordering $\leqslant$ over the elements in $\Gamma \cup \Gamma'$, if:

- $s = Eli$, then $\Gamma \unlhd_{Eli} \Gamma'$ if $\exists X \in \Gamma$ s.t. $\forall Y \in \Gamma', X \leqslant Y$;
- $s = Dem$, then $\Gamma \unlhd_{Dem} \Gamma'$ if $\forall X \in \Gamma, \exists Y \in \Gamma', X \leqslant Y$;

and we let $\Gamma \lhd_s \Gamma'$ iff $\Gamma \unlhd_s \Gamma'$ and $\Gamma' \ntrianglelefteq_s \Gamma$.

For any argument A, let $Prem_p(A) = Prem(A) \cap \mathcal{K}$, and $LastDefRules(A) = \emptyset$ if $DefRules(A) = \emptyset$, or $LastDefRules(A) = Conc(A_1), \ldots, Conc(A_n) \Rightarrow \psi$ if $A = A_1, \ldots, A_n \Rightarrow \psi$, otherwise $LastDefRules(A) = LastDefRules(A_1) \cup \ldots \cup LastDefRules(A_n)$. Then the *last-link* and *weakest-link* principles can be defined as follows.

Definition 2.7 [Last-/Weakest-link Principles] Let $s \in \{Eli, Dem\}$. Argument $A \preceq B$

- under the **last-link** principle, iff: 1. $LastDefRules(A) \unlhd_s LastDefRules(B)$; or 2. $LastDefRules(A) = \emptyset$, $LastDefRules(B) = \emptyset$, and $Prem_p(A) \unlhd_s Prem_p(B)$;
- under the **weakest-link** principle, iff:
  1. if $DefRules(A) = \emptyset$, $DefRules(B) = \emptyset$, then $Prem_p(A) \unlhd_s Prem_p(B)$; else
  2. if $Prem_p(A) = \emptyset$, $Prem_p(B) = \emptyset$, then $DefRules(A) \unlhd_s DefRules(B)$; else
  3. $Prem_p(A) \unlhd_s Prem_p(B)$ and $DefRules(A) \unlhd_s DefRules(B)$;

and we let $A \prec B$ iff $A \preceq B$ and $B \npreceq A$.

In addition to obtaining preferences over arguments based on preorderings on $\mathcal{K}_p$ and $\mathcal{R}_d$, CAS intends to take participants' priority orderings over values into account. We define the preordering $\leqslant$ based on $\lesssim$ defined in Definition 2.1 as follows.

Definition 2.8 [Priority Ordering Based on Values] Let $(\mathscr{L},^-,\mathscr{R},\mathscr{C},n)$ be a CAS and $C_i = \langle I_i, N_i, V_i, val, P_i \rangle$ be a context in $\mathscr{C}$. According to a preordering given by party x, denoted by $\lesssim_x \in P_i$, $\leqslant_x^v$ is a corresponding ordering on $\mathcal{N}_i$ s.t. $\forall n, n' \in \mathcal{N}_i, n \leqslant_x^v n'$ iff $val(n) \lesssim val(n')$.

Since in a context-based argumentation theory, the set of norms and values are explicitly specified, we have reason to believe that, arguably, if an argument does not include any norms or values, then it is constructed purely for epistemic reasoning, otherwise it is for practical (or decision-making) reasoning, or for a mixed purpose. On the one hand, a fundamental idea of CAS is that we emphasize argumentation always take place under certain social background, therefore participants share the same culture and beliefs, while disagreements arise due to the divergence on priorities over values. On the other hand, arguments for epistemic reasoning are intuitively in preference to arguments for decision-making or practical reasoning, since beliefs should be justified before they support any decisions. In conclusion, we can compare the epistemic elements in arguments first, and compare the other elements associated with values afterwards.

Let $\unlhd^v$ denote the ordering on sets of values related to each constructed argument, one can choose to compare two sets based on either Elitist or Democratic approach. Accordingly, we present the following definition for the context-based *structured ar-*

gumentation frameworks (SAF).

Definition 2.9 [Context-based Argumentation Frameworks] Let $CAT = (CAS, \mathcal{K})$ be an argumentation theory and $C_i = \langle I_i, N_i, V_i, val, P_i \rangle$ be a certain context, $\forall \lesssim_j \in P_i$, a structured argumentation framework $SAF_{C_{i-j}}$ is a tuple $\langle \mathscr{A}_i, CO_i, \preceq_i, \trianglelefteq_j^v \rangle$, where

- $\mathscr{A}_i$ is a set of arguments constructed from $\mathcal{K}$ based on CAT and C_i;
- $CO_i \subseteq \mathscr{A}_i \times \mathscr{A}_i$ is a set of conflict relations, $(A, B) \in CO_i$ iff A attacks B;
- $\preceq_i$ is an ordering on $\mathscr{A}_i$ based on two preorderings $\leqslant$ on $\mathcal{K}_p$ and $\leqslant'$ on $\mathscr{R}_d$;
- $\trianglelefteq_j^v$ is an ordering on $\{Values(X) | X \in \mathscr{A}\}$.

Based on the underlying idea of the current paper, $\leqslant$ and $\leqslant'$ can be understood as two priority orderings on the set of premises and the set of defeasible rules according to the (common) beliefs of all participants involved in the argumentation.

An alternative option is to compare the set of last norms or the set of all the norms according to $\leqslant^v$ on $\mathcal{N}_i$, corresponds to the last-link principle and the weakest-link principle. What worth a mention is that, as discussed in some literature (e.g. [23]), the last-link principle might lead to more appropriate outcome for normative reasoning, while the weakest-link principle might lead to more intuitive outcome for epistemic reasoning.

A Context-based Argumentation Framework can be seen as an extension of $ASPIC^+$ framework introduced in [22], while each $SAF_{C_{i-j}}$ constructed according to the preference of a certain party under context $C_i \in \mathscr{C}$ can correspond to a regular SAF of $ASPIC^+$ defined in [22], we call it a *standard SAF*.

Proposition 2.10 *A standard $SAF = \langle \mathscr{A}, CO, \preceq \rangle$ constructed based on an argumentation theory of $ASPIC^+$ is a special case of $SAF_{C_{i-x}}$ under context C_i, s.t. $\forall A \in \mathscr{A}_i$, $Norms(A) = \emptyset$.*

Proof. A standard $SAF = \langle \mathscr{A}, CO, \preceq \rangle$ is constructed from $\mathcal{K}$ in an argumentation system $AS = (\mathscr{L}, ^-, \mathscr{R}, n)$. [22]

Let $C_i = \langle I_i, N_i, V_i, val, P_i \rangle$ be a context, for each party x involved in the dispute, $SAF_{C_{i-x}} = \langle \mathscr{A}_i, CO_i, \preceq_i, \trianglelefteq_x^v \rangle$ based on $CAS = (\mathscr{L}, ^-, \mathscr{R}, \mathscr{C}, n)$ and $\mathcal{K}$. According to Definition 2.2, $\mathscr{R} = \mathscr{R}_s \cup \mathscr{R}_d \cup \mathcal{N}_i$. According to Definition 2.4, when constructing arguments, we apply norms in $\mathcal{N}_i$ in the same way as applying defeasible rules, therefore arguments in $\mathscr{A}_i$ are constructed just like $\mathscr{A}$ in a standard SAF; according to Definition 2.5, the set CO_i is obtained the same as for a standard SAF; as for the preferences $\preceq$ over arguments, we compare arguments based on $\trianglelefteq_x^v$ after they are compared based on $\preceq_i$ to get a more specific preference on arguments. $\forall A \in \mathscr{A}_i$, if no norm is applied, i.e. $Norms(A) = \emptyset$, therefore $Values(A) = \emptyset$, then $SAF_{C_{i-x}}$ can be reduced to $\langle \mathscr{A}_i, CO_i, \preceq_i \rangle$, which is the same as a standard SAF of $ASPIC^+$. $\square$

Given a structured argumentation framework, a set of defeat relations can be defined. In addition, we define a *value-based defeat* relation according to both $\preceq$ and $\trianglelefteq^v$. Let D denote the set of regular defeat relations and D^v denote the set of value-based defeat relations, the defeat relation according to CAS is defined as follows.

Definition 2.11 [Defeat] Let A, B be two arguments and $B' \in Sub(B)$,

(i) $(A,B) \in D$ iff A preference-independently attacks (i.e. undercuts, contrary-rebuts, or contrary-undermines) B on B', or A preference-dependently attacks B on B' and $A \not\prec B'$;

(ii) $(A,B) \in D^v$ iff $\{(A,B),(B,A)\} \subseteq D$ and $A \not\prec^v B$.

Whether an argument is acceptable can be evaluated based on Dung-style *abstract argumentation framework* (AAF) [15], which is defined as a tuple consisting of a set of arguments and a set of attack relations among arguments. We define an AAF based on the set of arguments obtained in CAT and the set of value-based defeats as follows.

Definition 2.12 [AAF] Let $SAF_{C_{i-j}} = \langle \mathscr{A}_i, CO_i, \preceq_i, \trianglelefteq_j^v \rangle$ be a structured argumentation framework and D_j^v be the set of defeats according to $\preceq_i$ and $\trianglelefteq_j^v$, an AAF (denoted by $F_{C_{i-j}}$) is a tuple $\langle \mathscr{A}_i, D_j^v \rangle$, and $\mathscr{F}_{C_i} = \langle \mathscr{A}_i, \mathscr{D}_i^v \rangle$ denotes a series of AAFs in context C_i, where $\mathscr{D}_i^v = \{D_1^v, D_2^v, \ldots, D_n^v\}$ ('1,…,n' represent all the parties in the disputes).

Based on an AAF, arguments are evaluated according to *argumentation semantics* [15]. With respect to certain argumentation semantics, a set of arguments that can be collectively accepted is called an *extension*. The following definition gives some standard argumentation semantics according to [15].

Definition 2.13 [Argumentation Semantics] Let $F_{C_{i-j}} = \langle \mathscr{A}_i, D_j^v \rangle$ be an AAF. An extension $E \subseteq \mathscr{A}_i$ is *conflict-free* iff $\nexists A, B \in E$ s.t. $(A,B) \in D_j^v$; A is *defended* by E (or *acceptable* w.r.t. E), iff $\forall B \in \mathscr{A}_i$, if $(B,A) \in D_j^v$, then $\exists X \in E$ s.t. $(X,B) \in D_j^v$, then:

- E is *admissible* iff E is conflict-free and each argument in E is defended by E;
- E is a *complete* extension iff E is admissible, and $\forall A \in \mathscr{A}_i$ defended by E, $A \in E$;
- E is a *grounded* extension iff E is the minimal [5] complete extension;
- E is a *preferred* extension iff E is the maximal complete extension.

2.2 Consensus based on CAS

Given certain argumentation semantic $S \in \{Complete, Grounded, Preferred\}$, let $\mathscr{E}_S$ denote the set of all the extensions for an AAF w.r.t. S. We define a consensus between two parties involved in a dispute under context C_i as follows.

Definition 2.14 [S-consensus] Let $\mathscr{F}_{C_i} = \langle \mathscr{A}_i, \mathscr{D}_i^v \rangle$ be a sequence of AAFs under context C_i, '1,…,n' represent all the parties in the dispute [6]

(i) $\forall F_{C_{i-j}}, F_{C_{i-k}} (1 \leqslant j,k \leqslant n$ and $j \neq k)$ in the sequence, if $\exists E, E'$ s.t. $E \in \mathscr{E}_{Si-j}$, $E' \in \mathscr{E}_{Si-k}$, and $E = E'$, then we say $O^{j,k} = \{Conc(A)|A \in E\}$ is a S-consensus among parties j and k;

(ii) if $\exists E \in \mathscr{E}_{Si-j}$, $\forall F_{C_{i-k}}$, there $\exists E' \in \mathscr{E}_{Si-k}$ s.t. $E = E'$, then we say $O_i = \{Conc(A)|A \in E\}$ is a S-consensus for all parties under context C_i.

In the previous paper [29], we hope to define a consensus as the maximal common subset of acceptable extensions under argumentation semantics S for each of the participant. Based on this idea, the (i) and (ii) of Definition 2.14 is as follows:

---

[5] 'minimal/maximal': both w.r.t. set-inclusion.

[6] those who give $\preceq_1, \ldots, \preceq_n \in P_i$ respectively

(i) $\forall F_{C_{i-j}}, F_{C_{i-k}} (1 \leqslant j, k \leqslant n$ and $j \neq k)$ in the sequence, if $\exists E, E'$ s.t. $E \in \mathscr{E}_{Si-j}$, $E' \in \mathscr{E}_{Si-k}$, and $E \cap E' = E$, then we say $O^{j,k} = \{Conc(A)|A \in E\}$ is a S-consensus among parties j and k;

(ii) if $\exists E \in \mathscr{E}_{Si-j}$, $\forall F_{C_{i-k}}$, there always $\exists E' \in \mathscr{E}_{Si-k}$ s.t. $E \cap E' = E$, then we say $O_i = \{Conc(A)|A \in E\}$ is a S-consensus for all parties under context C_i.

However, the subset of E' may not be an admissible set. So we modify the definition to guarantee a consensus is always a set of conclusion corresponding to a collectively acceptable set of arguments for each participant.

According to Definition 2.14, we can only get consensus that are collectively acceptable to each participants. Nevertheless, some consensus may be meaningless, for example, we may get a empty set. This raises the question: *What kind of consensus do we hope to achieve?*

In Definition 2.1, we have defined every context with a set of issues, which contains the controversial topics or decisions/actions that are under consideration. Through CAS, we hope the outcome can at least provide some certain answers to the issues, preferably provide decisions on all the issues.

Let $Cl_\neg(S)$ denote the closure of S under negation, we define the 'proper' and 'desired' consensus as follows.

Definition 2.15 [Proper & Desired Consensuses] Let O be a consensus among certain parties in a dispute under context $C_i = \langle I_i, N_i, V_i, val, P_i \rangle$, we say O is *proper* for a context C_i, iff $O \cap I_i \neq \emptyset$; we say O is *desired* for C_i, iff $Cl_\neg(O) = I_i$.

Correspondingly, O is *proper* for all the context $\mathscr{C} = \{C_1, \ldots, C_n\}$ of a CAS, iff $O \cap \mathscr{I}_i \neq \emptyset$, and O is *desired* for $\mathscr{C}$, iff $Cl_\neg(O) = \mathscr{I}_i$.

2.3 A Case Modelling

To illustrate how to get consensus through CAS, we introduce an example adapted from a historical event that took place in Ming dynasty of China:

The 'Great Rites Controversy' In 1521, the 14-year-old Jiajing Emperor succeeded the throne from his first cousin, the Zhengde Emperor, after the latter died childless. In order to perform the proper rituals according to some traditional documents and obey the 'Clan law', the Grand Secretary Yang Tinghe and many officials insisted that it was necessary that the Jiajing Emperor be posthumously adopted by his late uncle. However, Jiajing Emperor preferred to grant his own late father the title of Emperor. He and his supporters argued that they must obey the 'Filial piety' according to some other traditional documents.

We use 't' and 't'' to denote two different 'traditional documents' Jiajing and Yang respectively refer to, 'g' to denote 'grant Jiajing's biological father', 'v_{fp}' to denote the value 'Filial piety', and 'v_{cl}' to denote the value 'Clan law'. The first main context in this event can be modelled as follows according to Definition 2.1.

Context 1 of the 'Great Rites Controversy': $C_1 = \langle I_1, N_1, V_1, val, P_1 \rangle$, where
$$I_1 = \mathscr{I}_1 = \{g, \neg g\}, N_1 = \mathscr{N}_1 = \{n_1 : t \xrightarrow{v_{fp}} g; n_2 : t' \xrightarrow{v_{cl}} \neg g\},$$
$$V_1 = \{v_{fp}, v_{cl}\}, val(n_1) = v_{fp}, val(n_2) = v_{cl}, P_1 = \{\lesssim_1, \lesssim_2\}$$

Jiajing's ordering 1 (according to $\lesssim_1$) : $v_{cl} < v_{fp}$
Yang's ordering 1 (according to $\lesssim_2$) : $v_{fp} < v_{cl}$

In the history, the 'Great Rites Controversy' lasts for three years. Two parties represented by Jiajing and Yang had been arguing about whether to grant his father until Jiajing announced that if he could not grant his father, then he would abdicate. The context had changed and Yang's party made a concession. Because of loyalty, one of the most important values ancient ministers need to follow, they did not want the dynasty to become turbulent.

Let 'abd' denote 'Jiajing abdicate', 'tur' denote 'the dynasty to be turbulent' and 'v_{lo}' denote the value 'loyalty', the second main context of this event can be modelled as follows according to Definition 2.1.

Context 2 of the 'Great Rites Controversy': $C_2 = \langle I_2, N_2, V_2, val, P_2 \rangle$, where
$I_2 = \{abd, \neg abd\}$ and $\mathscr{I}_2 = \{g, \neg g, abd, \neg abd\}$,
$N_2 = \{n_3 : \neg g \xrightarrow{v_{cl}} abd; n_4 : \neg tur \xrightarrow{v_{lo}} \neg abd\}$ and
$\mathscr{N}_2 = \{n_1 : t \xrightarrow{v_{fp}} g; n_2 : t' \xrightarrow{v_{cl}} \neg g; n_3 : \neg g \xrightarrow{v_{cl}} abd; n_4 : \neg tur \xrightarrow{v_{lo}} \neg abd\}$,
$V_2 = \{v_{fp}, v_{cl}, v_{lo}\}$, $val(n_1) = v_{fp}$, $val(n_2) = v_{cl}$, $val(n_3) = v_{cl}$, $val(n_4) = v_{lo}$,
$P_2 = \{\lesssim'_1, \lesssim'_2\}$
Jiajing's ordering 2 (according to $\lesssim'_1$) : $v_{cl} < v_{fp}$
Yang's ordering 2 (according to $\lesssim'_2$) : $v_{fp} < v_{lo}, v_{cl} < v_{lo}$

The following arguments can be constructed based on CAS and $\mathscr{K} = \mathscr{K}_p = \{t, t', \neg tur\}$ (A, B denote arguments constructed under C_1, C_2 respectively):

$A_1 : t$ $A_2 : t'$ $A_3 : A_1 \xrightarrow{v_{fp}} g$ $A_4 : A_2 \xrightarrow{v_{cl}} \neg g$

$B_1 : \neg tur$ $B_2 : B_1 \xrightarrow{v_{lo}} \neg abd$ $B_3 : A_4 \xrightarrow{v_{fp}} abd$

Under the first context, according to Definition 2.11, $D_1^v = \{(A_3, A_4)\}$, $D_2^v = \{(A_4, A_3)\}$, consequently, no common extension exists for Jiajing's and Yang's parties after argument evaluation under any standard argumentation semantics. Under the second context, $D_1'^v = \{(A_3, A_4), (B_2, B_3), (B_3, B_2)\}$, $D_2'^v = \{(A_3, A_4), (A_4, A_3), (B_2, B_3)\}$, after argument evaluation, there is a common extension $\{A_1, A_2, A_3, B_1, B_2\}$ w.r.t. every standard argumentation semantics, and the corresponding set of conclusion is $\{t, t', g, \neg tur, \neg abd\}$, which is a desired consensus between Jiajing's party and Yang's party for all the contexts in $\mathscr{C}$.

2.4 Basic Rationality Postulates

In [10] Caminada and Amgoud declare four *rationality postulates* that any rule-based argumentation formalisms should at least fulfill, which are *sub-argument closure, closure under strict rules, direct consistency,* and *indirect consistency.*

According to Proposition 2.10, $SAF_{C_{i-x}}$ constructed according to a $CAT = (CAS, \mathscr{K})$ and a context C_i is the extension of standard SAFs of $ASPIC^+$. Therefore, adapted from [22], we define a 'well-defined' $SAF_{C_{i-x}}$ for party x as follows.

Definition 2.16 [Well defined $SAF_{C_{i-x}}$] Let $CAT = (CAS, \mathscr{K})$ be a context based argumentation theory, where $CAS = (\mathscr{L}, ^-, \mathscr{R}, \mathscr{C}, n)$. We say that an $SAF_{C_{i-x}}$ is well-defined, if it is

- closed under contraposition or transposition, i.e. iff either:
  (i) for all $Q \subseteq \mathscr{L}$ and $\varphi \in Q$, $\psi \in \mathscr{L}$, if $Q \vdash \psi$, then $Q \setminus \{\varphi\} \cup \{-\psi\} \vdash -\varphi$; or
  (ii) if $\varphi_1, \ldots, \varphi_n \to \psi \in \mathscr{R}_s$, then for each $i = 1 \ldots n$, there is
  $\varphi_1, \ldots, \varphi_{i-1}, -\psi, \varphi_{i+1}, \ldots \varphi_n \to -\varphi_i \in \mathscr{R}_s$;
- axiom consistent: iff $\nexists \varphi, \psi \in Cl_{R_s}(\mathscr{K}_n)$ s.t. $\varphi = -\psi$;
- c-classical [7] : iff for any minimal set Q s.t. $\exists \varphi$, $Q \vdash \varphi, -\varphi$, it holds that $\forall \varphi \in Q$,
  $Q \setminus \{\varphi\} \vdash -\varphi$;
- well formed: if whenever $\varphi = -\psi$, then $\psi \notin \mathscr{K}_n$ and ψ is not the consequent of a strict rule.

As it has been proved that a well-defined standard *SAF* of *ASPIC*$^+$ satisfies the four rationality postulates [22], follow Proposition 2.10, the following proposition is straightforward.

Proposition 2.17 *Let E be a complete extension of a well-defined SAF$_{C_{i-x}}$, then*

- $\forall A \in E$, *if* $A' \in Sub(A)$, *then* $A' \in E$;
- $\{Conc(A)|A \in E\} = Cl_{\mathscr{R}_s}(\{Conc(A)|A \in E\})$;
- $\nexists \varphi, \psi \in \{Conc(A)|A \in E\}$ *s.t.* $\varphi \in \overline{\psi}$;
- $\nexists \varphi, \psi \in Cl_{\mathscr{R}_s}(\{Conc(A)|A \in E\})$ *s.t.* $\varphi \in \overline{\psi}$.

Proposition 2.17 specified that a well-defined *SAF$_{C_{i-x}}$* satisfies the basic rationality postulates [10] under all the complete argumentation semantics.

3 Comparisons and Related Work

3.1 Comparisons with Aggregation Approaches

Aggregation methods from *social choice theory* [4,9] are commonly used to obtain collective decisions in multi-agent argumentation systems [12,13,21]. For argumentation formalisms involving norms or values (e.g. [5,6,7,19,20]), there are three possible approaches to aggregate opinions: aggregating the priorities on values [21], aggregating the AAFs [13,21], and aggregating the alternative extensions[12].

Approach Based on Preferences Aggregation Aggregation of value priorities can be realised based on the technique of preferences aggregation. Briefly, by this method different orderings on values will be aggregated by an aggregation function. Consequently, a collective preference is obtained, based on which a collective AAF can be constructed and extensions of arguments can be obtained after argument evaluation. To apply preferences aggregation method, participants have to reveal their preferences, in other words, explicit contextual information need to be given. Therefore, it has the closest idea to the current paper compared with the other aggregation methods.

To compare with preference aggregation method, for simplicity, we assume the priority orderings over V_i in CAS is a linear order (i.e. an irreflexive, transitive and complete binary relation).

---

[7] 'c' stands for 'contradictory'

Then we take a classical aggregation rule, the *Borda count* [27], as an example. For m values in the set V, the most preferred value to a party x can get $m-1$ points from x, the second preferred value can get $m-2$ points from x, and so on until the last preferred value gets 0 points.

Consider the first context of the Example introduced in Section 2.3. Assuming there exist a third party in the dispute, for instance, a people's representative, who believed an Emperor should prioritize the value of 'Clan law' over the value of 'Filial piety'. We can get the following priorities.

Jiajing Emperor:	$v_{cl} < v_{fp}$
Yang Tinghe:	$v_{fp} < v_{cl}$
People's Representative:	$v_{fp} < v_{cl}$

The scores based on Borda count are v_{cl} : 2 and v_{fp} : 1. So a collective preference on v_{cl} and v_{fp} is $v_{fp} < v_{cl}$. Although a conclusion can be drawn, Jiajing Emperor's preference would be violated.

Approach Based on Graph and Judgement Aggregation Based on AAFs, graph aggregation technique is to aggregate and obtain a collective set of attacks (defeats). A graph aggregation can be achieved without explicitly specifying the context, or in other words, the priority orderings over values. However, it is possible that a output of a collective graph is against every participants' preferences [21].

Judgement aggregation deals with the obtained extensions through AAFs, aiming to get a collective result based on all the alternative extensions. Although every candidate extension is justified, it is possible that a consequent result is not an admissible set of arguments [13].

Summary As analysed above, the preferences aggregation approach can guarantee the justification of the consensus, but probably violates some participants' interests. The graph or judgement aggregation approach can leave the contextual information implicit. However, for the former, it cannot guarantee the justification of the consensus, and it is possible to draw conclusions that every participant is not satisfied with; for the latter, it may lead to a result not collectively acceptable to some of the participants.

Through a CAS, the consensus as an output is both justified and in line with every participants' opinion. Under some contexts, there may be no consensus exists, then in order to reach some consensus, at leat one party has to shift the context, that is, propose new norms and values to change the other participants' priority orderings over values. Therefore, instead of drawing a result within the divergence based on some principle (such as the simple majority rule), CAS supports to keep the dialogue advancing until some consensus emerges.

From a pragmatic perspective, CAS can reflect some common strategies in debates, as we can see from the running example adapted from a real historical event. What is more, this setting is in line with the dynamic nature of argumentation.

3.2 Value-based Argumentation Framework

In [5,6,7], Bench-Capon et al. introduce a Value-based Argumentation Frameworks (VAF), which is an extension the abstract argumentation framework [15]. It compares arguments and get defeat relations based on the values each argument promotes. Many researches that consider multi-agent reasoning and values are developed on the basis of VAF and combined with ideas from social choice theory, such as [1,26].

VAF restricts that each argument can be associated with only one value. If only this case is considered, CAS is like an instantiation of VAF that includes argument structures. In this sense, CAS can also be seen as a combination of $ASPIC^+$ and VAF.

3.3 Some Other Related Work

Booth et al.[8] present a property-based argumentation framework based on the *preference-based AF* [2] and *property-based preference* [14]. Given that values can be regarded as a property, it associates each argument with a set of properties it satisfies, and models the dynamics of preferences by the change of "motivational states" within a dialogue.

For argumentation systems including contextual information, Amgoud et al.[3] propose an argumentation framework based on contextual preferences, in which an ordering on contexts in defined and one can select the "best" context. Ways for disagreement solution it suggests are mainly based on aggregation approaches. There are researches built on other argumentation formalisms (e.g. Assumption-based Argumentation and Defeasible Logic Programming) that consider contextual factors [28,30].

For normative reasoning and values, Liao et al.[20] present an argumentation theory for ethical practical reasoning includes norms and values based on $ASPIC^+$. Kaci et al. [19] propose approaches to compare arguments associated with multiple values.

As for the notion of consensus, Possebom [25] designed a mechanism to calculate the consensus for decision making using argumentation, which took the strength of arguments into account, while the current paper proposes a qualitative definition of consensus based on the existing argumentation semantics.

4 Conclusions

Based on the Generalized Argumentation Theory [16,17,18] and formal argumentation theory, especially $ASPIC^+$ [22], this paper reconsiders the rather preliminary design ideas of the Context-based Argumentation System (CAS) proposed in a previous paper [29]. Some definitions are modified to make them more in line with the pragmatic features of real argumentation, as well as more reasonable. Meanwhile, a more detailed discussion of how to get consensus based on CAS is given.

Compared with other argumentation formalisms, CAS 1) highlights the dynamic feature of contexts, 2) can deal with normative reasoning involving multi-agents, 3) proposes a way to obtain consensus combining the pragmatic perspective. What is more, based on a CAS, the acceptable conclusions for each parties in the dispute under each context C_i are clear, which is promising to provide easy-to-understand clues for explaining the obtained consensus.

References

[1] Airiau, S., E. Bonzon, U. Endriss, N. Maudet and J. Rossit, *Rationalisation of profiles of abstract argumentation frameworks: Characterisation and complexity*, Journal of Artificial Intelligence

Research **60** (2017), pp. 149–177.

[2] Amgoud, L. and C. Cayrol, *A reasoning model based on the production of acceptable arguments*, Annals of Mathematics and Artificial Intelligence **34** (2002), pp. 197–215.

[3] Amgoud, L., S. Parsons and L. Perrussel, *An argumentation framework based on contextual preferences*, in: *Proceedings of the international conference on formal and applied and practical reasoning*, 2000, pp. 59–67.

[4] Arrow, K. J., "Social choice and individual values," Wiley: New York, 1951.

[5] Bench-Capon, T., K. Atkinson and A. Chorley, *Persuasion and value in legal argument*, Journal of Logic and Computation **15** (2005), pp. 1075–1097.

[6] Bench-Capon, T. and P. Dunne, *Value based argumentation frameworks*, in: *Proceedings of 9th International Workshop on Non-Monotonic Reasoning*, 2002, pp. 444–453.

[7] Bench-Capon, T. J. M., *Persuasion in Practical Argument Using Value-based Argumentation Frameworks*, Journal of Logic and Computation **13** (2003), pp. 429–448.

[8] Booth, R., S. Kaci and T. Rienstra, *Property-based preferences in abstract argumentation*, in: P. Perny, M. Pirlot and A. Tsoukiàs, editors, *Algorithmic Decision Theory* (2013), pp. 86–100.

[9] Brandt, F., V. Conitzer, U. Endriss, J. Lang and A. D. Procaccia, editors, "Handbook of Computational Social Choice," Cambridge University Press, 2016.

[10] Caminada, M. and L. Amgoud, *On the evaluation of argumentation formalisms*, Artificial Intelligence **171** (2007), pp. 286–310.

[11] Cayrol, C., V. Royer and C. Saurel, *Management of preferences in assumption-based reasoning*, in: B. Bouchon-Meunier, L. Valverde and R. R. Yager, editors, *IPMU '92—Advanced Methods in Artificial Intelligence* (1992), pp. 13–22.

[12] Chen, W. and U. Endriss, *Aggregating alternative extensions of abstract argumentation frameworks: Preservation results for quota rules.*, in: *Proceedings of the 7th International Conference on Computational Models of Argument, COMMA*, 2018, pp. 425–436.

[13] Chen, W. and U. Endriss, *Preservation of semantic properties in collective argumentation: The case of aggregating abstract argumentation frameworks*, Artificial Intelligence **269** (2019), pp. 27–48.

[14] Dietrich, F. and C. List, *Where do preferences come from?*, International Journal of Game Theory **42** (2013), pp. 613–637.

[15] Dung, P. M., *On the acceptability of arguments and its fundamental role in nonmonotonic reasoning, logic programming and n-person games*, Artificial Intelligence **77** (1995), pp. 321 – 357.

[16] Ju, S., *The theory and method of generalized argumentation*, Social Sciences in China **XXXI** (2010), pp. 73–89.

[17] Ju, S., *The theory and method of generalized argumentation*, Studies in Logic **13** (2020), pp. 1–27.

[18] Ju, S., W. Liu and Z. Chen, *L'argumentation sur la titulature impériale dans la dynastie ming au prisme de la "théorie généralisée de l'argumentation"*, Argumentation et Analyse du Discours **25** (2020), pp. 1–23.

[19] Kaci, S. and L. van der Torre, *Preference-based argumentation: Arguments supporting multiple values*, International Journal of Approximate Reasoning **48** (2008), pp. 730–751.

[20] Liao, B., M. Slavkovik and L. W. N. van der Torre, *Building jiminy cricket: An architecture for moral agreements among stakeholders*, in: *Proceedings of the 2019 AAAI/ACM Conference on AI, Ethics, and Society, AIES 2019*, Honolulu, HI, USA, 2019, pp. 147–153.

[21] Lisowski, G., S. Doutre and U. Grandi, *Aggregation in value-based argumentation frameworks*, in: L. S. Moss, editor, *Proceedings Seventeenth Conference on Theoretical Aspects of Rationality and Knowledge, TARK 2019, Toulouse, France, 17-19 July 2019*, EPTCS **297**, 2019, pp. 313–331.

[22] Modgil, S. and H. Prakken, *A general account of argumentation with preferences*, Artificial Intelligence **195** (2013), pp. 361–397.

[23] Modgil, S. and H. Prakken, *The aspic+ framework for structured argumentation: a tutorial*, Argument & Computation **5** (2014), pp. 31–62.

[24] Perelman, C. and L. Olbrechts-Tyteca, "The New Rhetoric: A Treatise on Argumentation," University of Notre Dame Press, Notre Dame, Indiana, 1969.

[25] Possebom, A. T., *Consensus decision-making using argumentation*, in: *Proceedings of the International Joint Conference on Autonomous Agents and Multiagent Systems, AAMAS 2017*, Sao Paulo, Brazil, 2017, pp. 1853–1854.

[26] Pu, F., J. Luo, Y. Zhang and G. Luo, *Social welfare semantics for value-based argumentation framework*, in: M. Wang, editor, *Knowledge Science, Engineering and Management* (2013), pp. 76–88.

[27] Saari, D. G., *The borda dictionary*, Social Choice and Welfare **7** (1990), pp. 279–317.

[28] Teze, J. C., S. Gottifredi, A. J. García and G. R. Simari, *Improving argumentation-based recommender systems through context-adaptable selection criteria*, Expert Systems with Applications **42** (2015), pp. 8243 – 8258.

[29] Yu, Z., *A context-based argumentation framework with values*, in: *Proceedings of the 20th Workshop on Computational Models of Natural Argument, CMNA 2020*, Perugia, Italy (and online), 2020, pp. 1–10.

[30] Zeng, Z., X. Fan, C. Miao, C. Leung, J. J. Chin and Y. S. Ong, *Context-based and explainable decision making with argumentation*, in: *Proceedings of the 17th International Conference on Autonomous Agents and Multiagent Systems, AAMAS 2018* (2018), pp. 1114–1122.

Analysing plausible reasoning with a gradual argumentation model

Bin Wei[1]

Guanghua Law school of Zhejiang University

Abstract

The plausible reasoning deduces the plausible conclusion from the plausible premises or a plausible inference. The plausibility is different from probability, which makes it inappropriate to apply probability calculation for plausibility reasoning. The formalization of plausible reasoning needs to be discussed in the context of argumentation to characterize the dynamics of the plausibility. The gradual argumentation model provides a calculation for evaluating the dynamics of the plausibility of plausible arguments and progressively defines three plausibility standards: apparent plausibility, validated plausibility, and stable plausibility standards.

Keywords: plausible reasoning, gradual argumentation model, $ASPIC+$

1 Introduction

The concept of plausibility can be traced back to the Greek skeptics and Sophists. In argumentation theorist's view, plausible reasoning is regarded as the third kind of reasoning other than deductive and inductive reasoning. The plausible inference implies that if the premises are true, the conclusion is plausibly true. In this sense, a plausible proposition is a provisionally acceptable hypothesis because it seems correct, and there is no reason to think it is wrong. In other words, a proposition that is plausible often means that it seems or appears to be true, or if it fits in with other things people accept as true. The plausibility is seen as the appearance of such acceptability, where the connection is not constant but only appears to be true.

Intuitively, plausible reasoning often relies on people's perception of the appearance of things. For example, something that seems to have an apparent property is the *prima facie* reason to support that thing with some property, that is, the reason makes people accept that the proposition is true. In the view of informal logicians, the implication of plausibility means that it puts a weight of support behind a proposition, which gives a basis for accepting that proposition tentatively, where there is a reason for choosing between accepting it or not, or between accepting or rejecting it[12].

---

[1] Email:srsysj@zju.edu.cn. This work is supported by the Major Program of National Social Science Fund (20ZD047).

Plausible reasoning has also attracted great interest from mathematicians and logicians, G.Polya believed that probability can be used to express plausibility, and Bayes' theorem is applicable to plausible reasoning[7]. N.Rescher adopted another way of expressing plausibility, using the degree of plausibility to measure plausibility[10]. However, none of these works studied the dynamics of plausibility: how to evaluate the change in the plausibility of a conclusion when new premises appear to affect the conclusion in different ways. Informal logicians have recognized the research on plausible reasoning in the view of argumentation theory. This paper aims to apply a gradual argumentation model to analyse and evaluate plausible reasoning.

2 Plausible reasoning in formal argumentation

The formalization of plausible reasoning needs to satisfy specific rules, which are often derived from the characteristics of plausible reasoning. These characteristics include at least: (1) The plausibility of the conclusion should not be weaker than the premise with the weakest plausibility. (2) Plausible reasoning is defeasible, and it can be tested in context of argumentation or dialogue. (3) Plausible reasoning can be quantified, but pascalian probability calculus is not applicable. (4) There exist standards for testing plausibility, among which stability is an important standard. The most controversial focus among these features is why plausible reasoning is not suitable for probability calculus, which requires more detailed discussion.

Plausibility is usually considered similar to probability. The degree of plausibility of a proposition can be calculated. The calculation is a function of the initial probative force of the evidence supporting it minus the probative force of any of the contrary indicators that may have been introduced by the testing of the probability[11]. However, the plausible inference is inherently different from probable inference. The probability calculus requires that the sum of the probability of P and non-P should be equal to 1, but in plausibility calculus, both a proposition and its negative proposition can be highly plausible, the sum of plausibility value is not necessarily equal to 1. In the ancient example of plausible inference, the negation rule will not work. Therefore, we argue that plausibility reasoning does not apply to probability theory but applies other quantitative methods of plausibility.

As a new non-monotonic formal system, formal argumentation puts the inconsistencies in the reasoning system in a directed graph composed of a set of arguments and a set of attacking relationships. The abstract argumentation semantic proposed by Dung in 1995 is a pioneering work in this field[2]. Abstract semantics reflect the different kinds of acceptability of arguments by defining some argument sets containing the relationship of attacking. Formal argumentation models have been extended with stronger properties, such as $ASPIC+$ framework[4,8].

However, argumentation semantics in terms of preferences can not express the gradual nature of practical argumentation. For example, in the dialogue of judicial trials, the judge often needs to give the judgment based on the strengths of the arguments of the prosecutor and the defender. The strength of an argument is strong or weak, not limited in comparing which is better or worse. Therefore, the preference-based argumentation model can not assess the plausibility of arguments. The plausibility value of the premise supporting the conclusion is progressive, and a specific plausible argument is strongly plausible or weakly plausible. To solve this problem, we proposed a gradual argumentation semantic[1] based on $ASPIC+$, which critically examines Pollock's critical-link semantic with variable degrees of justification[6].

The advantages of studying plausible reasoning with the gradual argumentation model include: (1) the structured argumentation framework can clearly show how the premise or conclusion of a plausible argument is attacked. (2) the gradual argumentation semantics can evaluate the dynamics of plausibility after being attacked by other arguments. (3) the gradual argumentation model can also provide a quantitative method for evaluating the plausibility in addition to the probability calculus.

3 Standards of plausibility

In Carneades' theory,there are three grades of plausibility in an ascending order—some *phantasiai* are (1)just plausible, some are (2)plausible and tested, and some are (3)plausible, tested, and stable[12]. New academices prefers the plausible and tested *phantasia* to the simply plausible, and to both of them the *phantasia* that is plausible, tested and stable[3].

Following this view, the standard of plausibility can be divided into three levels. The first standard is apparent plausibility. It refers to the initial evaluation of the plausibility of the plausible argument without considering all attacking arguments. The second type is validated plausibility. It is considered that the initial plausibility of the plausible argument is weakened or attacked by other arguments, but still possesses a certain degree of plausibility. The third type is stable plausibility, which means that the plausibility has reached a stable state, and its plausibility will not be influenced by other counter-arguments and will not change significantly.

Therefore, the first step of the plausible standard is to define the apparent plausibility, which needs to meet the weakest plausibility value (the threshold is α). Only when this standard is met, can an argument be said to be plausible. Intuitively, if the plausibility value $Pr(A)$ of A is greater than the threshold α , which represents the lowest plausibility, then A is plausible.

Definition 3.1 [apparent plausibility] Argument A is apparently plausible if and only if $Pr(A) > \alpha > 0$.

Plausible argument A is validated plausible means that in addition to meeting the standard of apparent plausibility, the revised plausibility value $Pl(A)$ of A must be positive, that is to say, the initial plausibility value of A must be greater than all attacking arguments.

Definition 3.2 [validated plausibility] Argument A is validated plausible if and only if A is plausible and $Pl(A) > 0$.

Plausibility argument A is stably plausible means that in addition to meeting the validated plausibility standard, the revised plausibility value of A must not be less than the highest threshold β, and all attacking arguments do not meet the apparent plausibility standard.

Definition 3.3 [stable plausibility] Argument A is stable plausible if and only if A meets the validated plausibility standard, and $Pl(A) \geq \beta$, $Pr(B_i) < \alpha$, where B_i represents A's attacking argument.

4 Conclusion

This paper applies a gradual argumentation model based on $ASPIC+$ to give a formalization of plausible reasoning. This model provides an interpretable AI model for the practice of plausible reasoning, making the dynamic process of plausibility. The structured argumentation framework constructs an analysis theory and analyzes the structure of plausible arguments. The gradual argumentation semantic captures the dynamic changes of the plausibility of plausibility arguments, thus forming the evaluation theory of plausible arguments. In the future, we will explore more formal properties of calculation used to evaluate plausibility in the gradual argumentation model and will try to apply this model to real cases, such as legal cases.

References

[1] Wei,B. and H.Prakken. An Analysis of Critical-link Semantics with Variable Degrees of Justification[J]. Journal of Argument and Computation, 2016, 4:35-53.

[2] Dung,P.M. On the Acceptability of Arguments and its Fundmental Role in Nonmonotonic Reasoning, Logic Programming and N-person Games[J]. Artificial Intelligence, 1995, 77(2):321-357.

[3] Mates,B. The Skeptic Way: Sextus Empiricus' Outlines of Pyrrhonism[M]. New York: Oxford University Press,1996, p.122.

[4] Modgil,S.J. and H.Prakken. A General Account of Argumentation with Preferences[J]. Artificial Intelligence, 2013, 195:361-397.

[5] Pollock,J.L. Cognitive Carpentry. A Blueprint for How to Build a Person[M]. Cambridge:MIT Press, 1995.

[6] Pollock,J.L. Defeasible reasoning with variable degrees of justification, *Artificial Intelligence*, 133: 233-282, 2002.

[7] Pólya,G. Mathematics and Plausible Reasoning. Vol.II. Patterns of Plausible Inference[M]. Princeton University Press Princeton N J, 1968, 50(272).

[8] Prakken,H. An Abstract Framework for Argumentation with Structured Arguments[J]. Argument and Computation, 2010, 3(1):93-124.

[9] Renon,L.V. Aristotle's Endoxa and Plausible Argumentation[J]. Argumentation,1998, 12:95-113.

[10] Rescher,N. Plausible Reasoning: An Introduction to the Theory and Practice of Plausibilistic Inference[M]. Assen:Van Gorcum, 1976, p.15.

[11] Twining,W. Theories of Evidence: Bentham and Wigmore, London: Weidenfeld and Nicolson, 1985, p.55.

[12] Walton,D. Legal Argumentation and Evidence[M]. University Park, Penn State Press, 2002.

A formal approach to case comparison in case-based reasoning: research abstract

Heng Zheng[1]

Artificial Intelligence, Bernoulli Institute, University of Groningen
The Netherlands

Davide Grossi

Artificial Intelligence, Bernoulli Institute, University of Groningen
The Netherlands

ILLC/ACLE, University of Amsterdam
The Netherlands

Bart Verheij

Artificial Intelligence, Bernoulli Institute, University of Groningen
The Netherlands

1 Introduction

In this abstract, we introduce an approach about the comparison of cases in case-based reasoning with a formal theory that described in a series of research [2,3,5,6].

As we discussed in [6], our approach provides a new generalization and a new refinement of comparisons in case-based reasoning. We illustrate these contributions with an example (shown in Figure 1) from the domain of trade secret law of the United States, which has been discussed in [1,3,6]. As shown in Figure 1, in this example, the *American Precision* case[2] and the *Yokana* case[3] are considered as precedents, and the *Mason* case[4] is considered as a current situation, of which the outcome needs to be decided.

2 Method

We use a propositional logic language L generated from a set of propositional constants. We write $\neg$ for negation, $\wedge$ for conjunction, $\vee$ for disjunction, $\leftrightarrow$

---

[1] This paper is a research abstract of [5,6].
Corresponding Author: Heng Zheng, University of Groningen, Nijenborgh 9, 9747 AG Groningen, The Netherlands; E-mail: h.zheng@rug.nl.

[2] American Precision Vibrator Co. v. National Air Vibrator Co., 764 S.W.2d 274 (Tex.App.-Houston [1st Dist.] 1988)

[3] Midland-Ross Corp. v. Yokana, 293 F.2d 411 (3rd Cir.1961)

[4] Mason v. Jack Daniel Distillery, 518 So.2d 130 (Ala.Civ.App.1987)

A formal approach to case comparison in case-based reasoning: research abstract

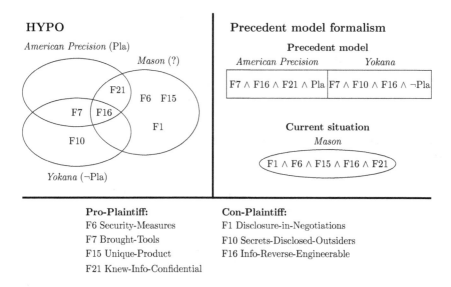

Pro-Plaintiff:
F6 Security-Measures
F7 Brought-Tools
F15 Unique-Product
F21 Knew-Info-Confidential

Con-Plaintiff:
F1 Disclosure-in-Negotiations
F10 Secrets-Disclosed-Outsiders
F16 Info-Reverse-Engineerable

Fig. 1. A Venn diagram [1] and a precedent model [3] about the *Mason* problem

for equivalence, $\top$ for a tautology, and $\bot$ for a contradiction. The associated classical, deductive, monotonic consequence relation is denoted $\vDash$.

Precedents consist of factors and outcomes. We consider both *factors* and *outcomes* are literals. A literal is either a propositional constant or its negation. We use $F \subseteq L$ to represent a set of factors, $O \subseteq L$ to represent a set of outcomes. The sets F and O are disjoint and consist only of literals. If a propositional constant p is in F (or O), then $\neg p$ is also in F (respectively in O). A factor represents an element of a case, namely a factual circumstance. Its negation describes the opposite fact. An outcome always favors a side in the precedent, its negation favors the opposite side.

Definition 2.1 [Precedents] A *precedent* is a logically consistent conjunction of distinct factors and outcomes $\pi = \varphi_0 \wedge \varphi_1 \wedge \ldots \wedge \varphi_m \wedge \omega_0 \wedge \omega_1 \wedge \ldots \wedge \omega_{n-1}$, where m and n are non-negative integers. We say that $\varphi_0, \varphi_1, ..., \varphi_m$ are the *factors* of π, $\omega_0, \omega_1, ..., \omega_{n-1}$ are the *outcomes* of π. If $n = 0$, then we say that π is a *situation* with no outcomes, otherwise π is a *proper precedent*.

Notice that both m and n can be equal to 0. When $m = 0$, there is one single factor. When $n = 0$, the precedent has no outcome and the empty conjunction $\omega_0 \wedge \ldots \wedge \omega_{n-1}$ is equivalent to $\top$. We do not assume precedents are complete descriptions. That is, factors may exist which do not occur in the precedent. Furthermore, we do not assume that the negation of a factor holds when the factor does not occur in the precedent.

Example 2.2 As shown in Figure 1, the precedents in the formalism are represented as follows:

(i) *American Precision*: $F7 \wedge F16 \wedge F21 \wedge Pla$;
(ii) *Yokana*: $F7 \wedge F10 \wedge F16 \wedge \neg Pla$;

(iii) *Mason*: F1 ∧ F6 ∧ F15 ∧ F16 ∧ F21.

A *precedent model* is a set of logically incompatible precedents forming a total preorder representing a preference relation among the precedents.

Definition 2.3 [Precedent models] A *precedent model* is a pair $(P, \geq)$ where P is a set of precedents such that for all $\pi, \pi' \in P$ with $\pi \neq \pi'$, $\pi \wedge \pi' \vDash \bot$; and $\geq$ is a total preorder over P.

As customary, the asymmetric part of $\geq$ is denoted $>$. The symmetric part of $\geq$ is denoted $\sim$.

Example 2.4 Figure 1 shows a precedent model with precedents *American Precision* and *Yokana*. As suggested by the size of the boxes, these two precedents are as preferred as each other.

Notions of comparing precedents in case-based reasoning include analogies, distinctions and relevances, they are related to general formulas, not only the factors or outcomes.

Definition 2.5 [Analogies, distinctions and relevances] Let $\pi, \pi' \in L$ be two precedents, we define:
(i) a sentence $\alpha \in L$ is an *analogy between π and π'* if and only if $\pi \vDash \alpha$ and $\pi' \vDash \alpha$.
(ii) a sentence $\delta \in L$ is a *distinction in π with respect to π' (π-π' distinction)* if and only if $\pi \vDash \delta$ and $\pi' \vDash \neg\delta$.
(iii) a sentence $\rho \in L$ is a *relevance in π with respect to π' (π-π' relevance)* if and only if $\pi \vDash \rho$, $\pi' \nvDash \rho$ and $\pi' \nvDash \neg\rho$.

Example 2.6 When comparing *Mason* with *Yokana* through the precedent model formalism:
(i) Analogies between *Yokana* and *Mason*: *e.g.*, F16, F16 ∨ F21, (F7 ∧ F10 ∧ F16 ∧ ¬Pla) ∨ (F1 ∧ F6 ∧ F15 ∧ F16 ∧ F21);
(ii) *Mason-Yokana* relevances: *e.g.*, F6 ∧ F15 ∧ F21, F1 ∧ F21;
(iii) *Yokana-Mason* relevances: *e.g.*, F10, F16 ∧ ¬Pla;
(iv) There is no distinction between *Mason* and *Yokana*.

3 Discussion and conclusion

The formalism we use for constructing precedent models is different from HYPO and CATO, as they describe cases as sets of factors. For instance, the *Yokana* case is represented by set {F7, F10, F16} in HYPO/CATO. While in our formalism, it is represented by a logical conjunction of factors and outcomes. Therefore, the comparison of cases in HYPO is by the notions related to sets, such as the relevant similarity (the set of shared factors by two cases, which is used for the reason that the two cases should have the same outcome) and the relevant difference (the set of unshared factors by two cases, which can be used for pointing out the two cases should be decided differently).

For instance, in the example shown in Figure 1, HYPO uses set {F16}, as the relevant similarity between *Mason* and *Yokana*, for the reason that *Mason* should have the same outcome as *Yokana*, and uses set {F6, F15, F21, F10} as the relevant difference between *Mason* and *Yokana* that can be used for arguing the two cases should have different outcome.

Comparing with our formalism, where we represent the relevant similarity and difference between *Mason* and *Yokana* as an analogy and a relevance respectively (Example 2.6), we thereby show that our approach provides a new generalization and a new refinement of the comparison in case-based reasoning. For the new generalization, our approach is able to not only compare cases by the factors themselves, but also compare them with the compound formulas that based on the factors, as shown in Example 2.6.

For the new refinement, our approach distinguishes the unshared formulas between cases as distinctions and relevances. As in Example 2.6, we can refine the relevant difference between *Mason* and *Yokana* with the relevances, and treat the different outcomes between *American Precision* (Pla) and *Yokana* (¬Pla) as distinctions.

The research abstract we present here shows a new generalization and a new refinement of case comparison in case-based reasoning with a formal theory. In recent publications, we further apply the approach to general case models [4], and discuss hard cases in law with the formalism by connecting the hardness with the involved arguments' validities [7]. The formal theory has the potential to further model case-based reasoning in the future.

References

[1] Ashley, K. D., "Artificial Intelligence and Legal Analytics: New Tools for Law Practice in the Digital Age," Cambridge University Press, Cambridge, 2017.

[2] Verheij, B., *Formalizing Arguments, Rules and Cases*, in: *Proceedings of the Sixteenth International Conference on Artificial Intelligence and Law*, ICAIL 2017, ACM, New York, 2017 pp. 199–208.

[3] Zheng, H., D. Grossi and B. Verheij, *Case-Based Reasoning with Precedent Models: Preliminary Report*, in: H. Prakken, S. Bistarelli, F. Santini and C. Taticchi, editors, *Computational Models of Argument. Proceedings of COMMA 2020*, Frontiers in Artificial Intelligence and Applications **326**, IOS Press, Amsterdam, 2020 pp. 443–450.

[4] Zheng, H., D. Grossi and B. Verheij, *Logical Comparison of Cases*, in: *Proceedings of the Artificial Intelligence and the Complexity of Legal Systems (AICOL) Workshop at JURIX 2020*, to appear, 2020 .

[5] Zheng, H., D. Grossi and B. Verheij, *Precedent Comparison in the Precedent Model Formalism: A Technical Note*, in: S. Villata, J. Harašta and P. Kšemen, editors, *Legal Knowledge and Information Systems. JURIX 2020: The Thirty-third Annual Conference*, Frontiers in Artificial Intelligence and Applications **334**, IOS Press, Amsterdam, 2020 pp. 259–262.

[6] Zheng, H., D. Grossi and B. Verheij, *Precedent Comparison in the Precedent Model Formalism: Theory and Application to Legal Cases*, in: *Proceedings of the EXplainable and Responsible AI in Law (XAILA) Workshop at JURIX 2020*, to appear, 2020 .

[7] Zheng, H., D. Grossi and B. Verheij, *Hardness of Case-Based Decisions: a Formal Theory*, in: *Proceedings of the Eighteenth International Conference on Artificial Intelligence and Law*, ICAIL 2021, to appear, New York, 2021 .

Formalizing the Right to Know:
Epistemic Rights as Normative Positions

Réka Markovich

University of Luxembourg

Olivier Roy

University of Bayreuth

Abstract

We argue for the fruitfulness of studying epistemic rights in the context of the theory of normative positions. We do so through an illustration considering the right to know. We show six possible formalizations of this right, and study their logical behaviors.

Keywords: Deontic and epistemic logics, dynamic logic, reasoning about knowledge.

What does it mean to say that expectant parents have a right to know whether their child will be healthy? Or that a patient has a right to know her test results? What do we refer to when saying that citizens have a right to know if their government does something illegal, or that a detained person has a right to know his rights?

These questions bear on epistemic rights. The category of epistemic rights has been studied in philosophy, but mainly regarding the right to believe in the context of epistemic justification [7]. In legal theory, epistemic rights have only been studied in contrast with normative positions [21,1]. It has even been argued that Hohfeldian categories (see below) cannot be used to analyze epistemic rights [22]. The systematic study of epistemic rights *as* a group of legal, Hohfeldian rights started recently [19,20]. The work presented here contributes to this literature, taking epistemic rights, in the narrow sense, to mean those rights that concern the epistemic state of the right-holder. In a broader sense, combinations of deontic and epistemic notions have been studied in cases where the rights bear on the duty-bearer's epistemic state, such as the right to privacy or the right to be forgotten [2,3,6]. An important benchmark here is the so-called Aqvist's paradox [16,10], which we will come back to later.

Here we focus on the right to know, and show that it can be fruitfully studied in the theory of normative positions. The right to know has many versions depending on what kind of knowledge the right concerns, and it's ranging over the different normative positions (examples below). There are other epistemic rights, of course: the freedom of thought, consumers' right not to be misled by advertisements, or the right to truth, some of which we address in the companion paper [12].

Normative Positions The theory of normative positions stems from the work of the American legal theorist, Hohfeld [9], who proposed a distinction between four types of atomic right-positions (Figure , top line) and their correlative duty positions (Figure , bottom line) [11]:

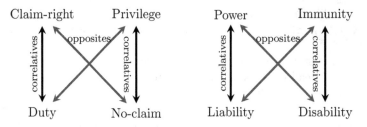

Each right corresponds to a correlative duty, and for the present paper, this correlation can be simply seen as equivalence. The normative positions are inherently relational. An agent's claim-right concerns the counter-party's action, and this constitutes the latter's duty (in the narrow sense). An agent's privilege (or freedom) to do something means that the counter-party doesn't have a claim-right that he doesn't do that thing. A power and the related positions in the second square are higher-order. Having a power means having the potential to change the counter-party's normative positions. Immunity refers to the counter-party's lack of power.

When we refer to someone's right in law, it can be any of the above-mentioned atomic positions (or a combination thereof [11]). In Hungary, for instance, citizens have a right to know the MPs' declarations of property, also that of the local representatives. But these two rights are different normative positions. MPs have a statutory duty to publish their declarations, while local representatives have to publish them only if a citizen requests it. The first right to know is a claim-right, while in the case of local representatives, it is a power. The right to know also comes under different guises in healthcare. If a medical test is carried out, the patient has a right to know the result: the doctor has a duty to let him know. A right to *not* know is usually also listed among the patient's rights: the patient has the ability to change the doctor's duty to let him know about his results, which is a power.

Language and Semantics Here we only sketch the formal model, the precise definitions will be provided in the presentation and the companion paper. We use a propositional language extended with four modalities. $\mathbf{K}_a\phi$ is the standard knowledge modality from epistemic logic, to be read as "agent a knows that ϕ". $\mathbf{O}_{a\to b}(\phi/\psi)$ is a directed conditional obligation, to be read as "given ψ, a has a duty towards b that ϕ". For a study of the importance of such directed obligations in the theory of normative positions, see [11]. $E_a\phi$ is an agency operator to be read as "agent a sees to it that ϕ", and $\Box\phi$ is a (legal) necessity operator to be read as "it is legally settled that ϕ".

This language is interpreted in Kripke models extended with a neighborhood function f_a for the agency operator. We assume that the epistemic accessibility relations are equivalence relations, that the conditional obligation operators are interpreted using the standard preferential semantics [8], and that the legal necessity operator is interpreted also on an equivalence relation. The neighborhood function is only assumed to satisfy the following: for all w and $X \in f_a(w)$, $w \in X$. Later on, we also show the consequences of assuming that the neighborhood functions are monotone, i.e. if $X \in f_a(w)$ and $X \subseteq Y$ then $Y \in f_a(w)$. Effectivity functions in coalition logic have, for instance, that property [15].

We also investigate formalization of the right to know using dynamic modalities [18,17] instead of static agency operators. This is motivated by the fact that the actions we study here are mostly epistemic actions, for instance *informing* a patient about her test result. The dynamic operators are of the form $[\mathcal{A}_d, a]$, where $\mathcal{A}_d$ is an action model for agent d, and a is an epistemic action in it. These modalities are then interpreted with the standard product update operation [18].

The right to know as a normative position Our running example is one of expectant parents who have a right to know whether their fetus will be a healthy child or whether it has a genetic disorder, illness, etc, which we refer to using a the propositional constant *ill*. As a first pass, this scenario can be captured by stating that the parents have an unconditional claim-right against the doctor to know whether the fetus is ill. We can formalize this in either a static or dynamic way.

$$\mathbf{O}_{d \to p} E_d(\mathbf{K}_p(ill) \vee \mathbf{K}_p(\neg ill)) \tag{1}$$

$$\mathbf{O}_{d \to p} \bigwedge_{a \in \mathcal{A}_d} [\mathcal{A}_d, a](\mathbf{K}_p(ill) \vee \mathbf{K}_p(\neg ill)) \tag{1d}$$

Instead of an unconditional claim-right, one could instead capture the right to know as a pair of conditional obligations. There are two options here, the first one using the primitive conditional obligations operator available in our language both in static and dynamic forms.

$$\mathbf{O}_{d \to p}(E_d \mathbf{K}_p(ill)/ill) \wedge \mathbf{O}_{d \to p}(E_d \mathbf{K}_p(\neg ill)/\neg ill) \tag{2}$$

$$\mathbf{O}_{d \to p}(\bigwedge_{a \in \mathcal{A}_d} [\mathcal{A}_d, a]\mathbf{K}_p(ill)/ill) \wedge \mathbf{O}_{d \to p}(\bigwedge_{a \in \mathcal{A}_d} [\mathcal{A}_d, a]\mathbf{K}_p(\neg ill)/\neg ill) \tag{2d}$$

An alternative formalization uses the classical "wide scope" approach [5]:

$$\mathbf{O}_{d \to p}(ill \to E_d \mathbf{K}_p(ill)) \wedge \mathbf{O}_{d \to p}(\neg ill \to E_d \mathbf{K}_p(\neg ill)) \tag{3}$$

$$\mathbf{O}_{d \to p}(ill \to \bigwedge_{a \in \mathcal{A}_d} [\mathcal{A}_d, a]\mathbf{K}_p(ill)) \wedge \mathbf{O}_{d \to p}(\neg ill \to \bigwedge_{a \in \mathcal{A}_d} [\mathcal{A}_d, a]\mathbf{K}_p(\neg ill)) \tag{3d}$$

Observations and Results We first look at the logical relationships between these different formalizations. Since there is no set relationships between the static agency operators and the epistemic actions, the static and the dynamic versions are mutually independent. On the static side, (2) entails (3), while (1) and (2) are logically independent, and so are (1) and (3). If, however, the neighborhood functions are monotone, then we get stronger relationships: (3) entails (1), but not the other way around, and so (2) entails (1). The picture is different on the dynamic side: (1d) and (3d) become logically equivalent, and both are logically strictly weaker than (i.e. are implied by) (2d).

Next, we consider Aqvist's paradox. It arises from the basic observation that in standard deontic logic, if someone has an obligation to know that something bad is the case and knowledge is veridical, then it ought to be that something bad is the case [16]. Our formalizations put an agency or a dynamic operator between the deontic and the epistemic operators, but this operator also validates T, so the question arises whether they also fall prey to the paradox. Formulating an epistemic claim-right as a right to know *whether* generally allows the paradox to be avoided [10]. This is also true for (1), also in monotone frames, and for its dynamic version (1d). For the right to know formulated using conditional obligations, the situation is more subtle. Regardless of whether they express epistemic rights, conditional obligations validate $O_{d\to p}(ill/ill)$ [23]. So we get trivially that $\mathbf{O}_{d\to p}(E_d\mathbf{K}_p(ill)/ill)$ entails $\mathbf{O}_{d\to p}(E_d\mathbf{K}_p(ill) \wedge ill/ill)$. This has little to do with our formalization of this epistemic right. It arises from this specific semantics of conditional obligations. (2), however, does *not* imply an unconditional obligation that the fetus is ill. Like the first, it turns out that the wide-scope formulations also avoid the paradox, although it is also well-known that they handle contrary-to-duty obligations poorly [13]. In the full paper, we also study whether the paradox re-appears when we consider propositions that are (legally) settled.

We finally observe that dynamic analysis allows the relation between duties bearing on epistemic actions, and possible epistemic conditions for these duties, to be studied in more detail. As usual in dynamic epistemic logic (DEL), our three dynamic versions can be translated into formulas without a dynamic operator using reduction axioms. These formulas make explicit references to the pre-conditions of the actions. This allows the study of the effect of imposing stronger, possibly epistemic preconditions, for instance, that an action of informing parents that their fetus is ill is only executable when the doctor knows that the relevant test results are positive. This would allow an interesting connection to be made with the theory of knowledge-based obligations, for instance [14].

Conclusion We have sketched how epistemic rights can be fruitfully analyzed by combining tools from the theory of normative positions and epistemic logic. We did this through an illustration considering the right to know. We have shown that even for this comparatively simple right, one can devise at least six possible formalizations, with different logical properties. We take this to be a first step towards a more comprehensive analysis. The relationship between what the holder of the correlative duty ought to do in the claim-right to know and the state of knowledge of the right-holder should be better understood, for instance.[1] Here we represented this as the duty-bearer, i.e. the doctor, having a duty to see to it that the parents know whether the fetus is ill. A different content of this claim-right would be that the doctor has a duty to make the information about the health of the fetus available to the parents. Since this idea comes close to a claim-right to knowability, it might be analyzable using tools from the logic of arbitrary announcements [4].

References

[1] Altschul, J., *Epistemic entitlement*, Internet Encyclopedia of Philosophy (2021).
URL https://iep.utm.edu/ep-en/#H6

[2] Aucher, G., G. Boella and L. Torre, *A dynamic logic for privacy compliance.*, Artificial Intelligence & Law **19** (2011), pp. 187 – 231.
URL   http://proxy.bnl.lu/login?url=http://search.ebscohost.com/login.aspx?
direct=true&db=iih&AN=66352839&site=ehost-live&scope=site

[3] Aucher, G., G. Boella and L. V. D. Torre, *Privacy policies with modal logic: the dynamic turn*, in: *Deontic Logic in Computer Science (DEON 2010)*, 2010, pp. 196–213.

[4] Balbiani, P., A. Baltag, H. Van Ditmarsch, A. Herzig, T. Hoshi and T. De Lima, *What can we achieve by arbitrary announcements? a dynamic take on fitch's knowability*, in: *Proceedings of the 11th conference on Theoretical aspects of rationality and knowledge*, 2007, pp. 42–51.

[5] Broome, J., *Wide or narrow scope?*, Mind **116** (2007), pp. 359–370.

[6] Cuppens, F. and R. Demolombe, *A deontic logic for reasoning about confidentiality*, in: M. A. Brown and J. Carmo, editors, *Deontic Logic, Agency and Normative Systems, DEON '96: Third International Workshop on Deontic Logic in Computer Science, Sesimbra, Portugal, 11-13 January 1996*, Workshops in Computing (1996), pp. 66–79.
URL https://doi.org/10.1007/978-1-4471-1488-8_4

[7] Dretske, F., *Entitlement: Epistemic rights without epistemic duties?*, Philosophy and Phenomenological Research **60** (2000), p. 591?606.

[8] Hansson, B., *Deontic logic and different levels of generality*, Theoria **36** (1970), pp. 241–248.

[9] Hohfeld, W. N., *Fundamental legal conceptions applied in judicial reasoning*, in: W. W. Cook, editor, *Fundamental Legal Conceptions Applied in Judicial Reasoning and Other Legal Essays*, New Haven : Yale University Press, 1923 pp. 23–64.

[10] Hulstijn, J., *Need to know: Questions and the paradox of epistemic obligation*, in: R. van der Meyden and L. W. N. van der Torre, editors, *Deontic Logic in Computer Science, 9th International Conference, DEON 2008, Luxembourg, July 15-18, 2008. Proceedings*, Lecture Notes in Computer Science **5076** (2008), pp. 125–139.
URL https://doi.org/10.1007/978-3-540-70525-3_11

[11] Markovich, R., *Understanding hohfeld and formalizing legal rights: the hohfeldian conceptions and their conditional consequences*, Studia Logica **108** (2020).
URL onlinefirst:https://doi.org/10.1007/s11225-019-09870-5

[12] Markovich, R. and O. Roy, *A logical analysis of freedom of thought*, Proceedings of the 15th International Conference on Deontic Logic and Normative System (DEON 2021) (2021).

[13] McNamara, P. and F. Van De Putte, *Deontic Logic*, in: E. N. Zalta, editor, *The Stanford Encyclopedia of Philosophy*, Metaphysics Research Lab, Stanford University, 2021, spring 2021 edition .

[14] Pacuit, E., R. Parikh and E. Cogan, *The logic of knowledge based obligation*, Synthese **149** (2006), pp. 311–341.

---

[1] We thank one of the anonymous reviewers of the LNGAI conference for pointing out this issue.

[15] Pauly, M., *A modal logic for coalitional power in games*, Journal of logic and computation **12** (2002), pp. 149–166.

[16] Åqvist, L., *Good samaritans, contrary-to-duty imperatives, and epistemic obligations*, Nous **1** (1967), pp. 361–379.
URL http://www.jstor.org/stable/2214624

[17] Van Benthem, J., "Logical dynamics of information and interaction," Cambridge University Press, 2011.

[18] Van Ditmarsch, H., W. van Der Hoek and B. Kooi, **337**, Springer Science & Business Media, 2007.

[19] Watson, L., *Systematic epistemic rights violations in the media: A brexit case study*, Social Epistemology **32** (2018), pp. 88–102.
URL https://doi.org/10.1080/02691728.2018.1440022

[20] Watson, L., *The right to know: Epistemic rights and why we need them*, manuscript presented at the Edinburg Legal Theory Group, 24 October 2019 (2019).

[21] Wenar, L., *Epistemic rights and legal rights*, Analysis **63** (2003), pp. 142–146.

[22] Wenar, L., *Rights*, in: E. N. Zalta, editor, *The Stanford Encyclopedia of Philosophy*, Metaphysics Research Lab, Stanford University, 2015, fall 2015 edition .

[23] Zvolenszky, Z., *Is a possible-worlds semantics of modality possible? A problem for Kratzer's semantics*, , **12**, 2002, pp. 339–358.